Hell on Belle Isle

Ordering Information

Additional copies of Hell on Belle Isle may be purchased for $17.95, plus $5 postage and handling, from Faded Banner Publications, P.O. Box 101, Bryan, Ohio 43506-0101, telephone 1-888-799-3787, www.fadedbanner.com E-mail address is don@fadedbanner.com Dealer inquiries are welcome.

Hell on Belle Isle

Diary of a Civil War POW
Revised Edition

Journal of Sgt. Jacob Osborn Coburn
Editing and narration by Don Allison

Faded Banner Publications
Bryan, Ohio

Hell on Belle Isle Revised Edition. Copyright ©1997 and 2022 by Donald L. Allison. All rights reserved. Printed in the United States of America. No part of this book may be used or reproduced in any manner whatsoever, including information storage and retrieval systems, without written permission except in the case of brief quotations embodied in critical articles or reviews. For information address Faded Banner Publications, Box 101, Bryan, OH 43506-0101. www.fadedbanner.com

Library of Congress Control Number: 2022906918

Hell on Belle Isle: Diary of a Civil War POW
Revised Edition

Journal of Jacob Osborn Coburn
Allison, Donald L.

ISBN 979-8-9861060-0-7

First Printing of Revised Edition

I dedicate this book in the memory of my mother-in-law, Corabelle Eutsler. She, along with Osborn Coburn, taught me the true meaning of courage.

Acknowledgments

For their help in preparing this book I owe special gratitude to my wife, Diane, for her indulgence, support, insight and keen proofreading skills; my son, Stuart, for his assistance on artwork and graphic design; my mother and father, Jimmie and Charles Allison, for inspiring and cultivating my love of history and writing; Richard Cooley for making me aware of the Osborn Coburn diary, and sharing his research knowledge; and Mark Grisier for editorial and marketing assistance.

Editing and proofreading the manuscript, and making valuable suggestions, were Corabelle Eutsler, Christopher Cullis, Linda Freed, Becky Huntington, Coby Boszor and Carol Sloan.

I also thank Jeff Yahraus and Phil Scott of the staff of the Williams County Public Library in Bryan, Ohio; curator Michael Winey, U.S. Army Military History Institute at Carlisle Barracks, for help in locating the institute's Belle Isle photographs; Larry Strayer and Richard Baumgartner for publication advice and assistance regarding photographs; Ken Turner for permission to copy his photograph of 1st Lt. Robert A. Moon; Mark Weldon for his help in locating photographs and information; John Sickles for the John G. Anderson photograph and help in locating photographs and information on the 6th Michigan Cavalry; the staff at the Valentine Riverside Museum in Richmond, Va., for assistance in research and touring Belle Isle; the Richmond National Cemetery and Richmond National Battlefield staff for research help; the Mecosta County,

Mich., Historical Society; the Colorado Historical Society, Denver. Colo., for use of the photograph of Benjamin F. Rockafellow; the staff at the Jackson, Mich., Public Library; and the countless others who shared their time and knowledge to help make this book possible.

For this revised edition I owe special thanks to two people who kindly shared information with me after reading the first edition of "Hell on Belle Isle." Sally Butler of Fort Wayne, Ind., shared transcripts of additional letters written by Coburn and additional information regarding his diary. The late Jim Wood of Remus, Mich., a Big Rapids, Mich., native and historian, contacted me and generously sent me a considerable amount of his own research revealing details of Osborn Coburn's life in Mecosta County, Mich.

Table of Contents

Preface..12

Chapter I..17
Steady Watch of the Sentinel

Chapter II...29
A Man of Many Talents

Chapter III..39
On to Richmond

Chapter IV..60
Look and Listen, Wish and Wonder

Chapter V..65
Let Hope Predominate

Chapter VI..92
O, Won't We Be a Happy Crowd

Chapter VII...96
O Dear What Shall We Do

Chapter VIII..101
No I Shall Not Die Here

Chapter IX..112
To Build Castles in the Air

Chapter X..119
It Does Require a Stout Heart

Table of Contents

Chapter XI..123
I Shall Yet Come Out All Right

Chapter XII...133
My Old Camp Complaint

Chapter XIII 141
Nearer to an End

Epilogue..155

Appendix...166

Notes and Bibliography...179

Index..189

About the Author...195

Maps and Illustrations

J. Osborn Coburn..16

1st Lt. Robert Moon..21

Charles Town, W.Va..23

Lt. Benjamin "Frank" Rockafellow.......................................25

Map of Ohio and Michigan...35

Mount Jackson, Virginia..49

Map of Virginia..51

Pvt. John G. Anderson...53

Brick warehouse prison..56

Belle Isle bridge..69

Belle Isle Prison Camp...71

Belle Isle...73

Map of Richmond, Virginia...78

Released Belle Isle Prisoners..82-85

Major Thomas Turner on Belle Isle....................................90

Belle Isle and Richmond...97

Tredegar Iron Works...116

Maps and Illustrations

Belle Isle prisoners..126

Belle Isle..139

Cynthia Thomas..149

Lillian and Grantie Thomas..150

Belle Isle graves...153

Empty Belle Isle prison..162

J. Osborn Coburn marker dedication 163-164

J. Osborn Coburn memorial marker...................................165

Author Don Allison...195

Preface

I didn't want to narrate this book.

The truth is I had to.

It was pure chance that I learned of Jacob Osborn Coburn's diary. A notice about the journal was tucked in the corner of a photocopy given me by my good friend Richard Cooley. Richard and I are co-authoring a history of the 38th Ohio Volunteer Infantry regiment, and the photocopy from a 1935 Bryan, Ohio, Democrat contained a small article about the 38th. But as I went to file the article I noticed the small announcement telling of a Civil War POW's journal reprinted in the nearby Edgerton Earth newspaper. The diary had been supplied to the Earth by the late Verta Emanuel.

My interest piqued, I searched out copies of the old newspaper. With the help of former Earth publisher Robert Swope I obtained the reprinted segment of the diary, which appeared in 19 installments. Unfortunately it wasn't the entire journal, just the portion dealing with Osborn Coburn's experience as a prisoner of war. But had this section not been transcribed the entire diary would have been lost to posterity, as the original apparently was destroyed in a house fire in the 1970s.

My heart ached for Osborn as his words bridged the expanse of time. I laughed as he laughed, observed as he observed, hoped as he hoped. Tears welled in my eyes as he reflected on home and loved ones, and the ever-growing chance that he may never see them again. Even after I put the journal down Osborn's words stayed with me.

Osborn's story needed to be told. I prepared excerpts from the diary as a feature series in The Bryan Times of November 1989. The series seemed to strike a chord with the public, and I received considerable feedback. But these selected segments merely touched the surface of what Osborn experienced. I had known from the first the full story deserved to be told, and to a wider audience than a newspaper could offer.

Thus was born the idea for this book.

For years it remained just that, an idea. I had so many questions about Osborn. Research proved him to be singularly elusive, frequently moving, leaving little record behind. Bit by bit, frustrating as the search might be, the puzzle of his life began coming together. Even so, the many dead ends and seemingly endless hours of research made the project an easy one to shelve.

More than anything I wondered what Osborn looked like. I had discovered no photographs, and his few surviving distant relatives knew of none that existed.

In the fall of 1993 my wife Diane and I celebrated our wedding anniversary with several days of touring craft and antique shops in southern Michigan. We were on our way home, westbound on U.S. 12 approaching our turn south toward Ohio on U.S. 127, when we saw a small antique store ahead. We agreed we were too tired to stop, but for some reason I felt compelled to pull into the parking lot anyway. Diane agreed to a quick look inside. We had found nothing of interest and were on our way to the door when a tintype photograph caught Diane's eye. To pass the time as she examined the image I returned to a display case near the cash register. To my surprise I spotted a Civil War CDV – carte-de-visite, or small calling card photograph – I had overlooked earlier, and I asked the storekeeper for a closer look.

I felt strangely drawn to the image, a waist-up view of a seated soldier in a cavalry shell jacket. I turned the photo over, and drew in a sharp breath of shock.

Inscribed on the reverse was the name Osborn Coburn. When I looked back at the image and saw the brass number 6, crossed cavalry sabers and the letter I on the soldier's cap – Osborn served in Company I of the 6th Michigan Cavalry – I realized the impossible had happened. It was really him.

I looked into Osborn's eyes, and I knew it was time to finish the book.

J. Osborn Coburn

Steady Watch
of the Sentinel

Hunger's dull ache gripped Sgt. Jacob Osborn Coburn. His weary mind drifted toward sleep, but intermittent raindrops in the dark nudged him back to consciousness.

Osborn's blanket was gone, lost in the fight. Now his wool pants and jacket soaked up the chill dampness of the ground.

His feet hurt. They were chafed by cavalry boots suited more for riding than the day's 28-mile forced march on roads winding up and down West Virginia's hills. Confederates on horseback had kept the column moving ahead of Union pursuers. Now those Rebels stood around the nearly 450 resting Yankees, guns in hand as they kept a close watch on their captives. Campfires at intervals circled the group, hissing and crackling in the rain. The flames cast a flickering light around the prisoners, but the fires were too far away for Osborn to feel the warmth.

It had not been a good Sabbath day.

When dawn's glow filtered through the trees the guards shouted at the captives, forcing them to their feet and into column on the roadway to resume the trek south. Later, in the diary he carried, Osborn Coburn began recording his observations as a prisoner of war.

Oct. 18, 1863 — Fight commenced at 7:00 a.m., at 9:09 a.m. the entire Federal command was marching under a strong guard. We had no breakfast and marched till 11 o'clock next day with nothing to eat, when we got one pound of beef without any salt.

Oct. 19 — Eating some beef half cooked on coals, without salt. We rested till 3:30 o'clock when we left the south bank of the north branch of Shenandoah, Banks' old battle field near Fort Royal[1] and moved across south branch.

Did not go to Strassburg[2] as was first intended, a Federal force being in the way, we learned. This p.m. we moved on way up the stream along the stream's rock path and camped in a narrow valley upon the bank of the Shenandoah between two ranges of mountains.

Before their capture Osborn and his comrades guarded Charles Town, nestled among West Virginia's hills eight miles southwest of Harpers Ferry at the northern end of the scenic Shenandoah Valley. Four years earlier the nation's attention had focused on Charles Town, the village where abolitionist John Brown was tried and hanged for treason after his attack on the federal armory at Harpers Ferry.

Company I of the 6th Michigan Cavalry — Osborn's

---

[1] Actually Front Royal, Va.
[2] Actually Strasburg, Va.

unit — had been at Charles Town only a week, posted there to check roaming parties of Confederate raiders. The Company I cavalrymen rode down from Harpers Ferry, where they had been stationed for several months on detached duty from Brig. Gen. George Armstrong Custer's Michigan Cavalry Brigade of the Army of the Potomac. The 6th Michigan troopers reinforced the cavalrymen of Company F, 2nd Maryland Potomac Home Brigade, whose captain was killed and ranks reduced from continual skirmishing with Rebel horsemen.

Also with the Charles Town force was Capt. Samuel Means' cavalry company of Union independent Virginia Rangers, and six companies of the 9th Maryland Infantry Regiment. Mustered into the service only two months earlier, the green 9th Maryland troops had never before seen battle.

The sharp crack of carbine fire outside Charles Town pierced the pre-dawn darkness Sunday, Oct. 18, 1863. Picket guards retreated toward the Union camps of the town, warning their comrades of Rebel cavalrymen encircling the village. The Federals scrambled from their beds, threw on hats and jackets, grabbed their weapons and formed for battle.

About 1,500 southerners under Brig. Gen. John Imboden surrounded the 450 bluecoats of the Charles Town garrison. The Confederates, moving out from Berryville at 2 that morning, had ridden the 15 or so miles northeast to Charles Town. "The surprise was complete," explained Imboden, "the enemy having no suspicion of our approach until I had the town completely surrounded."

A courier carrying Imboden's surrender demand sought out Col. B. D. Simpson of the 9th Maryland Infantry, who commanded the Charles Town Federals. Col. Simpson asked for an hour to consider the matter. Imboden offered five minutes. Simpson replied, "Take us if you can."

Under Simpson's orders the Union cavalry mounted up and scouted outside the village, finding artillery aimed down at the town from low hills to the north and south. Rebels held the surrounding high ground and their battle lines commanded the roads leading out of the village. The 6th Michigan's 1st Lt. Robert Moon, leading the 75 or so men of the cavalry force, charged the Confederates blocking the road to Harpers Ferry. Most of the northern horsemen were killed, wounded or captured in the impetuous attack, with only Moon and 17 troopers escaping to return to the village and report their discovery.

Simpson sent his infantry to the center of town where they holed up in the brick courthouse, jail and other nearby buildings. Anticipating attack, the defenders had earlier cut loopholes in the walls of the structures. A thick oak wall enclosed the courthouse yard, offering further protection.

"The enemy," wrote Simpson, "sent in another flag of truce to notify the women and children to leave the town. Before the bearer could turn around to find out the time allowed, they began shelling us from their battery on the north side of town."

As Imboden explained, "I immediately opened on the buildings with artillery at less than 200 yards and with

First Lt. Robert Moon. (Photograph courtesy of Ken Turner.)

half a dozen shells drove out the enemy into the streets." Blasted from their stronghold, the inexperienced Union infantrymen struggled to form battle lines in the narrow streets. Officers shouted orders above the crash and din of the shells. But the men panicked and ran, scrambling for whatever safety they could find. Most headed toward Harpers Ferry but found the way still blocked by the Confederates.

According to Imboden, at the first shots Col. Simpson and four other mounted officers turned and fled, their horses running the gauntlet through Rebel fire and galloping toward Harpers Ferry. Union accounts put a more positive face on Simpson's behavior, stating the colonel maintained his cool and by commands and threats attempted to rally his men, fleeing only after his companies had scattered.

"One volley was exchanged," Imboden reported, "when the enemy threw down his arms and surrendered unconditionally."

A handful of northern soldiers somehow managed to flee the attackers. But Osborn Coburn and 26 others from Company I – about half the company – were among 434 Union troops taken prisoner. Confederates killed two bluecoats and wounded three in the Charles Town struggle. Among the wounded was Company I Pvt. Thomas Neal, whose foot was shot away. Neal, captured only four months earlier and then paroled to return to his unit, would survive the injury.

The Confederates expected Federal reinforcements from Harpers Ferry to arrive any time, so they quickly gathered up the captives and sent them marching south

View of Charles Town. ("From Frank Leslie's "Illustrated Leaders and Battle Scenes of the Civil War.")

under guard. Accompanying the column were several captured Union ambulances and wagons bearing weapons, ammunition, medicines and clothing.

Prisoners and supplies were kept moving ahead of a rear guard of Confederate horsemen who retreated from hill to hill, skirmishing with pursuing Union forces. Leading the Union cavalry advance in the chase was Lt. Benjamin "Frank" Rockafellow of the 6th Michigan Cavalry's Company M. Rockafellow was a quartermaster at Harpers Ferry, the same position held by Coburn at Charles Town. With about 50 mounted troopers, most from the 1st Connecticut Cavalry as his own Company M was on picket duty, Rockafellow was sent south from Harpers Ferry after the fleeing southerners. Approximately 600 Union infantry cooperated with Rockafellow's horsemen in the chase.

Finally, two miles from Berryville, the Federals withdrew.

For their rich haul in prisoners and booty the Rebels suffered only three men killed and about 20 wounded. The Union forces giving chase lost six killed and at least 43 wounded.

"The movement," Gen. Robert E. Lee later wrote to Imboden, "was well conceived and executed in a manner that reflects great credit upon yourself and the officers and men of your command, to whom I desire to express my appreciation of the brave and valuable service they have rendered."

For Osborn Coburn that well executed movement meant a long walk south in the hands of the Rebels.

Lt. Benjamin "Frank" Rockafellow. (Courtesy, Colorado Historical Society, negative F-779.)

Oct. 20 — *Our route is yet up the narrow and interesting valley. How great the contrast between this and the lower valley. There a gentle undulating surface once in a fair state of cultivation, Bolivar Hts. and Bunker Hill rising perceptibly above the other points, and well watered by beautiful small brooks. Here the valley is not more than two miles wide at any place, often not one and one-half of a mile and then very much "cut up". We have eaten our one small piece of meat at noon. Now we are encamped upon a small plateau and preparations are making for a "repast". This is up to 3 o'clock.*

Contrary to our expectations we remained here throughout the day. Drew some more meat and bread and have plenty to eat. Our treatment on the whole is somewhat better than I expected but the walk is very severe. Weather has cleared away and is becoming quite pleasant.

Oct. 21 — *We awoke this morning with a clear sky overhead. Glorious "Old Sol" is spreading his golden light broadcast over the green capped mountains that seem entirely to surround us. The drooping branches of the cedars, the cone topped pines interspersed here and there with the chestnut gold tinted by frost's displeasure, the occasional chirping of a bird or whistle of a quail, aye, everything trying to make me believe I am not a prisoner of war. But the steady watch of the sentinel near me, the tramp of "Grayback" troops all around, assures me that it is too true and that escape is impossible.*

Well we have a long march ahead today and the order is to start. And march we did till we had passed on and up the narrow valley, through a gap in the mountains resting awhile upon the summit attained then passing down, down to the Shenandoah proper (the other being but a tributary) which we had to ford. A mile or two more and we were upon the "Staunton Pike" where the Shenandoah proper, presented itself to our view in all its beauty, grandeur, age and glory, though its best portion is marred by the effects of war. Distant mountain ridges look beautiful while the undulating surface of the vale is really picturesque.

A Man of Many Talents

Like the land surrounding him, Osborn Coburn felt the effects of war. He was now 32 years old. Less than four months before, on June 11, 1863 at Seneca Mills in Maryland, he was hurt in a skirmish with the Confederates.

That battle occurred when about 250 Rebel horsemen under the noted raider Major John Mosby crossed the Potomac River at daybreak. The southerners dashed up the Chesapeake & Ohio Canal, driving in patrols and attacking the 76 troopers of Company I who guarded the canal locks at Seneca. Osborn described the affair in a letter:

Editors Pioneer; ... There was but a single company of 76 men stationed here. Nearly one half of these was absent from camp, at night, on picket and patrol duty, so that we could seldom muster more than 40 fighting men at any one time.

Our pickets fired the alarm just at day light, on the morning of the 11th, and immediately "Boots and Saddles" was sounded. By the time we were in our saddles the rebs, about 200 strong, were close upon us. We retreated about one mile and across the Seneca bridge, where we turned and fought them briskly until our pistols and carbines were empty. The rebs had divided their forces and came upon us from three directions, and were endeavoring to cut off our retreat. We retired about two miles further and again formed in line for battle but the enemy was close upon us with superior forces, our numbers had become somewhat reduced, and we again moved on.

The enemy having jading horses, while most of ours were fresh, deeming the pursuit useless, turned, retreated over the field, burning our camp and stores as they passed, and recrossed the Potomac. The crossing was effected over a ford about one mile and a half below the mouth of the Seneca.

We lost four killed, one slightly wounded, and seventeen missing, who it is supposed were all taken prisoners, among whom was Daniel W. Dutcher, of Green, Mecosta county. The killed and wounded were all from other localities. Lieut. Moon, Sergt. Williams, and three men were absent at Washington, having left but an hour or two before the alarm was given.

The enemy left one Captain and one Lieutenant dead on the field. Citizens say there were three dead carried off, and traces were found indicating other casualties. Our boys behaved well, and have been highly complimented for the pluck they showed in this their "first fight." They are in good spirits, and would not hesitate to display still greater valor, if possible, whenever an opportunity is given them. I write this brief account hurriedly, that the minds of the people at Big Rapids may be relieved from any concern as to their friends. ... Nothing but the vigilance and promptness of Captain [Charles] Deane prevented our entire Company being taken, with all our stores and camp equipage.

Even though the company was chased from the locks, officials at Washington D.C. – only 30 miles away – agreed with Osborn's initial assessment to the *Pioneer* and concluded the unit gave a good performance. Official reports differed, though, from Osborn's initial count of the casualties, stating that Company I lost four men killed, one wounded and 17 captured, while killing two Confederate officers and wounding one southern trooper.

Although Osborn's service records mention nothing about the injury he sustained in the fight, in his diary he alludes to his recovery. A bout with diarrhea, common for Civil War soldiers, also took a toll on his health.

Osborn was an attorney at Big Rapids, Mich., when he volunteered for the army in September 1862. He was born at Warren, Ohio, on Oct. 2, 1831, the son of Minerva (Twadell) and John F. Coburn. He had two younger siblings, a brother, Edwin R., and sister, Laura E. His brother also served in the Civil War, as a private in the 44th Indiana Volunteer Infantry from 1861 to 1864.

The Coburn family moved from Trumbull County, Ohio, to DeKalb County, Ind., in March 1836. Four years later Osborn's mother died. The following year his father remarried, to Alzada Matilda Gay, who died in 1850, and in 1852 he married Betsey Anna Wilmot.

A shoemaker, Osborn's father served as county clerk and recorder. He was a Whig and later a Republican, and took an interest in religious and educational affairs. John Coburn proved to be a major influence in his son's life.

By 1851 Osborn and a partner, "Ladd" Thomas, were selling goods from a log structure in Butler, Ind. A year later, on June 27, 1852, he and Martha Thomas were married in Defiance, Ohio. His wife, born in 1834, was one of five children of the Rev. Heman and Ann (Henry) Thomas. Like Osborn, Martha had lost a parent early in life when her father died in 1840. In 1848 her mother was remarried, to Israel Williams, and the couple along with Martha, her younger brother, Hawley Thomas, and infant half brother, Alfred, moved to Hicksville, Ohio, near the Indiana line and not far from Osborn's DeKalb County home. Osborn affectionately referred to Martha as Mattie.

By the mid 1850s Osborn was employed as a surveyor helping to lay out the route of the Air Line Railroad across Northwest Ohio. He and Mattie's brother Hawley both worked with the civil engineer crews extending the rails westward from Toledo, through existing young towns in Ohio's Great Black Swamp, the last part of the state to be settled. Along with existing towns, several new towns were established along the railroad right-of-way, including Stryker and Edgerton.

According to entries in a diary he kept at the time, on Jan. 1, 1855, Osborn was rooming with the James Winter family in Bryan. Later that month he boarded with a Mrs. Serrells in Bryan.

Construction of the region's first rail line was an exciting time for these Ohio towns. Osborn often was on board the first passenger train to arrive in each village, an event celebrated with exuberant festivities.

When not surveying the rail route Osborn was active

in music circles. His diary mentions attending numerous dances and concerts, including a Bryan Musical Association concert on April 7, 1855. He also was a charter member of the Woodbury Musical Association of Edgerton, and became the association's choir director.

His diary mentions attending church and various religious meetings – in fact, missing a Sunday service was significant enough to warrant mention in his diary.

In January 1856 Osborn and Mattie purchased a lot on the north side of Vine Street in Edgerton, facing toward the Air Line rails, and began building a home on the property. That same month Osborn was commissioned a notary public for Williams County. By mid June the house was complete and they moved in.

Osborn performed various odd jobs in addition to his work with the Air Line Railroad, such as surveying a mill seat on Beaver Creek. After the railroad was completed through Edgerton his service was no longer needed and he was discharged from the Air Line on Sept. 1, 1856. At that point his odd jobs became his livelihood, and he kept busy designing buildings, continued to do occasional work with the railroad, and helped survey the new Cincinnati and Mackinaw Railroad line from Defiance north into Williams County.

To achieve a more permanent occupation Coburn turned to a new career, the law. He studied under attorneys Schuyler Blakeslee of Bryan, who later would be elected to the Ohio legislature, and J.M. Ashley of Toledo, an abolition leader who would become a Republican member of the U.S. House of Representatives during the Civil War.

In the spring of 1859 Mattie's brother Hawley visited Columbia Township in Jackson County, Michigan, where his mother and stepfather had settled. Hawley taught school there in the winter of 1859. But by July 1860 Hawley was back in Ohio, living with Osborn and Mattie in their modest Edgerton home.

Osborn and Mattie sold their property in Edgerton on Oct. 4, 1860, and later that month relocated to Jackson County to be with Mattie's mother and stepfather. Hawley joined the family migration to Michigan, and he taught school in Brooklyn in 1860-61.

On June 12, 1861, Osborn suffered a devastating loss when Mattie died at the home of her mother. She was buried in a small cemetery at the west end of Clark Lake in Jackson County.

After Mattie's passing Osborn again pulled up stakes and moved further north, to Big Rapids in Mecosta County, Mich. Like Edgerton and other towns rising from the Great Black Swamp of Northwest Ohio that he had left behind, Big Rapids was a new town, carved out among the trees of northern Michigan's wilderness. With the new location came his new occupation, as one of Big Rapids' earliest attorneys.

At the time no law degree was required to practice as a lawyer, but his study of law in Ohio had prepared him to pass the Michigan bar, and on April 15, 1862, Osborn was granted a Michigan law license. The May 22, 1862 issue of Big Rapids' newspaper, the *Mecosta County Pioneer*, contained an advertisement for J.O. Coburn as an attorney at law, solicitor, notary public, and real estate and general collecting agent. And he had wasted little time in becoming involved in his new

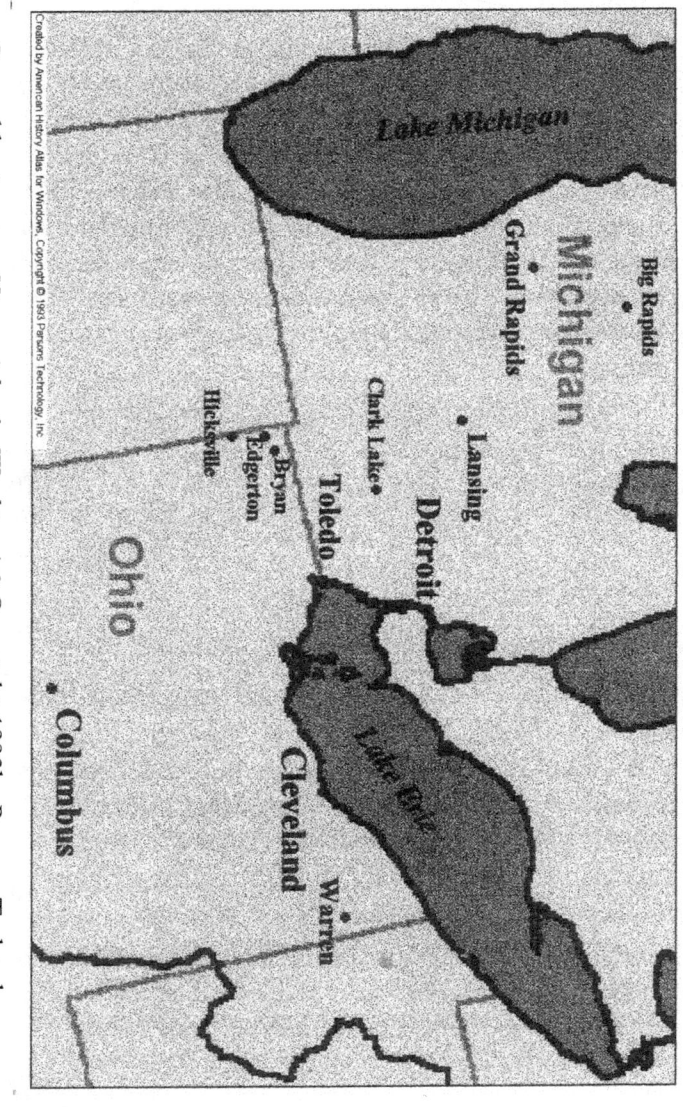

Created by American History Atlas for Windows 1.0 Copyright 1993 by Parsons Technology, Inc. Used by permission.

community, as he was elected clerk of Leonard Township on April 7. Too, he was appointed by the commissioner of the state land office as the Swamp Land Agent for the state.

Osborn established his law office with fellow attorney C.C. Fuller. Fuller, who doubled as the village postmaster, ran his law practice upstairs above the village post office, at the northwest corner of Elm and Michigan streets. By July 1862 Coburn and Fuller were law partners.

Osborn and Fuller were both 30 years old. Like Osborn's brother-in-law Hawley Thomas, Fuller had attended Hiram College in eastern Ohio. While at Hiram Fuller had been a roommate of future U.S. President James A. Garfield. Fuller, whose given name was Ceylon, had arrived in Big Rapids in May 1860.

Much like Osborn Coburn, Fuller was a man of varied talents and interests. Recognizing the value of a newspaper to a fledgling town like Big Rapids, Fuller enticed printer Charlie Gay to set up shop in the village, offering rent-free space below his law office for the printing press and agreeing to edit the paper with no compensation. The first issue of this paper, the *Mecosta County Pioneer*, was published April 17, 1862.

In addition to his work as an attorney and solicitor Fuller advertised his services as an insurance agent and circuit court commissioner. He attended to the payment of taxes and conducted a general agency business. Fuller also was appointed that year as probate judge.

Both Fuller and Coburn were involved in planning for Big Rapids' 1862 Fourth of July celebration. Coburn served as the event's marshal, and Fuller read the

Declaration of Independence. The two also served together on the Leonard Township Board of Inspection for schools.

In his reporting for the *Mecosta County Pioneer* Fuller was not above poking some fun at his law partner, referring to himself as "senior" partner even though they were the same age. He also kidded Coburn about an extended fishing trip, although he was grateful for the pickerel that Coburn and his fishing partner presented to him on their return.

Fuller also took notice in the newspaper of Coburn's increased interest in the affairs of the school — and hinted that the interest was more in a teacher than the classroom itself. That instructor, listed in township records as Rachel Aldrich, had begun teaching that spring term. Writing of his fellow school board of inspectors member in the May 22, 1862 *Pioneer*, Fuller noted that Coburn "is the best fitted of any of its members for the peculiar duties evolving upon a visitor of schools in an 'official capacity,' and accordingly appointed him a 'committee of one' to visit all the schools in this township as often as in his judgment he may deem it necessary. As we now have but one district in the township, and the School House is in the village, this will not be much of a tax upon his time, unless he chooses to make it so, as it is not usual to visit schools oftener than once a week."

Miss Aldrich returned to the classroom following the 1862 summer break. The *Pioneer* of July 24 noted that "The Fall term of our village school commenced on Monday last, under the charge of Miss A.R. Aldrich, who became popular as a teacher last term. ... We are proud of our school, and hope it may long continue under the charge of its accomplished teacher."

Osborn's courting of Miss Aldrich became more serious as the year progressed. Although variously referred to as Rachel or A.R., her given name was Everence, and Osborn referred to her as Eva. She and Osborn became engaged that year. At that time married women were not allowed to work as teachers, and later that fall Eva was replaced.

But the Civil War, now entering its second year, interrupted their romance, and Osborn left Eva behind when he volunteered for the army. Soon Eva returned to her native New York to reside with her parents, Thomas and Dulcena Aldrich, at Onstott.

On to Richmond

Until September 1862 Osborn Coburn had resisted Abraham Lincoln's earliest calls for troops. But when the 6th Michigan Cavalry was organized in late summer 1862, as a swell of patriotism swept the state, he enlisted in the new regiment. As a lawyer Osborn was likely swayed by the 6th Michigan's commander, noted attorney George Gray, a leader in the Western Michigan bar. Osborn knew Gray, who had been a fellow participant in the Big Rapids Fourth of July celebration that summer. A $50 enlistment bounty offered by Mecosta County was an added inducement.

When Coburn departed, C.C. Fuller took over his duties with Leonard Township, stepping in as deputy clerk.

The 6th Michigan's initial rendezvous was at Grand Rapids, Mich., where the men began the transition from civilians to troopers. With his close ties to the *Mecosta County Pioneer*, Osborn began serving as a correspondent in the ranks for the paper. From Camp Kellogg at Grand Rapids Osborn wrote Oct. 3 about the new soldiers' experiences:

Being relieved for a moment from constant labor during the day till "drill call" (3 o'clock P.M.,) I improve the opportunity to say a word for the Big Rapids people, now in camp. Camp Kellogg is again "all noise and confusion" – no, not confusion – for everything seems to move along quite smoothly, and in tolerable order. The furloughed soldiers have nearly all returned to camp, which swells the number to about 1,700 men, and still they come. The overplus, I am informed, are to be organized into the 7th Cavalry Regiment.

We have today been visited by the Mustering Officer, but for some reason the 6th is not to be mustered in till the coming week. Nearly all have received their uniforms, and the balance are to receive them as fast as they can be reached.

The boys from Big Rapids are all in fine spirits, and seem to enjoy camp life first rate. It is a new thing to them, and I fear that as the newness wears away, and the camp rules are more vigorously enforced, they will see less fun in it, and begin to realize that they are soldiers, not citizens. How the change, when fully felt, will relish, is yet to be learned. I think there are none but have come determined to do their duty, and conduct themselves as becomes "American Soldiers." To fight the battles of their country manfully, and when peace is again restored, to return home to their families and friends; as good, but wiser men than when they left.

Later that month Osborn provided the *Pioneer* with an update:

Editors Pioneer ...Yesterday was the day of days to the newly enlisted soldier, among the boys in Camp

Kellogg. Col. Smith, of the United States army, was here and mustered in the Regiment. But very few were thrown out, but among them we regret to mention the name of our gallant and very gentlemanly Adjutant, Patten. The Adjutant seemed to regret it very much, and we hope he may yet be assigned some place in the regiment. Lt. Col. Gray has been promoted to the Colonelcy of the Regiment and is very popular among the soldiers. I suppose the Regiment to be fully organized and ready for the service as soon as equipped, and shall have made sufficient proficiency in the drill and other exercises.

Perhaps a brief account of the method of "mustering in" will not be uninteresting. An entire company is called into line, upon the left of the officer, when the name of each, commencing with the Captain, and following in order of rank, is called, who pass as their names are called by the mustering officer, forming in another line to his right. A few horses are provided, and each in the order above mentioned is required to mount, trot about forty paces, and walk back. The entire company is then sworn. This completes the "mustering in." But, it of course is followed up by three cheers for the Colonel, and three for everybody else.

Osborn returned to Rig Rapids for a brief furlough Nov. 2. In December the regiment went by rail to Washington, D.C. There the men were issued sabers, revolvers and state-of-the art repeating Spencer rifles. Later the rifles were replaced with shorter Spencer carbines, better suited to horseback service.

From the regiment's camp in Washington on New

Year's Day 1863, Osborn wrote to the *Mecosta County Pioneer* newspaper:

The Tattoo is sounding, and the pale moon is lighting the soldier's way to his quarters, to answer to his name at the roll call. The bands are finishing their last piece of music for the night, and the air is cool and bracing. Many of the soldiers feel as happy as though in the "old house at home," though without a cent, while many wend their way slowly and sadly to their tents, their day's duty over, there to lie down wrapped in their blankets, to pass perhaps a sleepless night, in thinking of their destitute families at home, and themselves powerless to aid them; while one, your humble servant, sits in his cotton domicile indicting this communication for that unpretending, though worthy sheet, the Pioneer.

We should have commenced this by wishing you and your readers "a Happy New Year" and begging pardon for the oversight, ask all to receive it now, with this assurance, that the success of the Pioneer and its readers is ever uppermost on our mind.

We had expected to speak of the Capitol, but after reflection, have concluded that the subject is too big for us to handle in our limited space; suffice it to say, that it is the richest specimen of architecture we ever saw, and requires an abler pen than ours to do it justice, besides, being obliged to visit it with a "pass" in our fingers as evidence that we were not out "without leave," did not correspond well with our former notions of American liberty, and somewhat dampened the ardor with which, under more favorable circumstances, we should have viewed that stupendous pile, a fit monument to the wealth and liberality of a free and enlightened people.

This reminds us of an amusing incident. A sentinel of our regiment the other day was charged to let no one pass, except commissioned officers, and innocently asked if he should let the negro waiters pass. He was answered in the affirmative. When relieved two hours afterward, the relief asked him for the instructions, when he promptly replied, "Let no one pass but commissioned officers and niggers." Rather humiliating, thinks the American, that in the immediate vicinity of our National Capitol, the "nigger" has greater privileges than the white man. But all negroes are loyal, while too many whites are the opposite; hence we suppose this, to us, strange military necessity.

The Big Rapids boys are all well, unless it be some slight ailing. ... Don't fail to send the Pioneer to such of them as are subscribers, as it brings tidings of home, and is better than a day's rations.

During the winter the 6th Michigan took part in two raids across Long Bridge into Virginia, one involving a ride under the Englishman Sir Percy Wyndham to Falmouth, Va., where the Army of the Potomac was in camp. In February companies I and M – Osborn was in Company I – were detached to Poolesville, Md. In the spring of 1863 the regiment broke camp and rode to Fairfax Court House, Va., for picket duty.

At this point companies I and M were detached for service in the Shenandoah Valley, where the troopers skirmished several times with the Confederates. The Seneca Mills fight, where Coburn was hurt, was one such incident.

In an Aug. 9 letter to the *Pioneer*, Coburn detailed the movements of Company I following the June 11 battle. At this point Gen. Robert E. Lee's Army of Northern Virginia had begun its invasion of the North, and Union cavalry was probing to locate the Confederate forces.

June 24. – Commenced march to Maryland Heights, arrived June 26th, during a heavy rain, and in the mud.

June 27. – Marched to Sharpsburgh. Charged through the town, capturing 39 prisoners.

July 3. — Commenced march to Falling Waters arriving there at daylight on the 4th; had a slight skirmish with the rebs; destroyed rebel pontoon bridge there, and a small train of wagons and ammunition. Returned via Hagerstown to Frederick with 17 prisoners.

According to one account, during this July 4 movement – while the armies faced each other to the north at Gettysburg – the 6th Michigan's Companies I and M were part of a 140-man detachment that destroyed Lee's pontoon bridge over the Potomac River at Falling Waters. The northern horsemen captured the Confederate bridge guard, consisting of a lieutenant and 16 men, then destroyed pontoon boats with axes. After seizing three wagonloads of small arms ammunition, which they threw into the river, the troopers then returned to their base at Frederick, Md. With heavy rains raising the Potomac River beyond fording depth, the loss of the pontoon bridge delayed Lee's retreat from Gettysburg.

In his Aug. 9, from Harpers Ferry, Osborn continued his account of Company I's movements.

On the morning of the 7th inst., a cavalry force of one hundred men, with three days' rations, was ordered out in the direction of Leesburg, upon some mission not yet made public. Captain Deane being in command of all the cavalry in this brigade, the command of the expedition was assigned to Capt. Vinton, of Co. M, who proceeded across the Shenandoah, around the eastern point of Virginia Heights. Then dividing his command, he proceeded with a portion by the Berlin route on the Virginia side of the Potomac, while the balance, under Lieut. Moon, took the road through Hillsborough gap, both parties meeting in the vicinity of Waterford, in the latter part of the day. Captain Vinton then marched his entire command through the village, and learning that the guerrilla chief, White[3], with near 300 men, was but a mile or two away, selected an open space upon an eminence, to the eastward of the village, for his camping ground. The horses were fastened to a fence, while the men slept on their arms in the open field, which was mainly a gentle slope towards the village. Immediately back of the camping ground was a shelter of thick woods. Pickets were sent out at every probable place of approach. But the enemy was cunning, passing our pickets about twelve o'clock at night, under cover of the darkness, so skillfully that but a single shot was fired as

---

[3] Confederate Major Elijah V. White.

an alarm. In an instant our men were in their saddles, but the darkness was such, we could not tell the direction the enemy was coming from until he opened fire on us with a brisk fire from a dismounted line, and not more than 30 yards distant.

The reported superiority of their numbers induced Captain Vinton to order an immediate retreat, upon which Co. M and 40 of the 1st Conn. cavalry, who were with us, dashed quickly by. Co. I, not having heard the order did not execute it, but were left to shift for themselves, in rather an unpleasant manner. Lieut. Moon had his men in splendid line and was doing good execution in the way of fighting, having killed several of the enemy, wounding several others, among them one rebel captain, when he found himself with but 20 or 25 men among the enemy. He gave the order to retreat, and retired about a mile and a half to the eastward of the village on the road to Point of Rocks, where they awaited the return of day. So far as known the casualties to our men were one killed, two wounded, and eighteen missing.

We soldiers are generally, considerably worn down with excessive duties, but otherwise in the best of spirits.

It was in the Shenandoah Valley that Osborn, who began service as a corporal, was promoted to quartermaster sergeant on July 1, 1863. That was a natural assignment for Coburn, making use of his experience as Leonard Township clerk back at Big Rapids.

And he would serve in the Shenandoah Valley until

the October 1863 Charles Town fight.

As a captive following the Charles Town disaster, Osborn apparently marched south from Front Royal along Passage Creek, which flows northeast between the north and south forks of the Shenandoah River. Now, in the valley of the Shenandoah's north fork, his journey continued.

October 22 – The morning opens beautiful from our camp about one mile south of the little village of Mount Jackson. We are getting ready to move.

Three o'clock. We are resting, I am beneath the shade of the red cedars upon the bank of a clear running brook. The mountain range a few miles to the southward are but a shade or two deeper than the clear sky above. The country is very pretty and October has done much to improve it but that other "act of war" has in places made sad havoc of the architecture, mechanics and laborers effort. Our march along "The Pike" a well macadamized road is a very tedious one. We cannot get enough to eat. The citizens have not got it. One told a soldier, "If it had not been for our (your) d-----d army we could have had plenty."

Yankee replied, "If it hadn't been for their (your) d----d army ours would not have passed through here."

Secesh[4] replied, "Well, that's so," seeming to appreciate all that was said and meant.

We shall have to move soon. Wonder how soon our "days march" will end and how near is that to the end of

---

[4] Slang for secessionist, or Confederate sympathizer.

our march? Continued our march to within two miles of Harrisonburg, where we camped. I think I never was so footsore, fatigued and completely worried out as tonight.

October 23 ⁻ Feel much rested this morning and resumed the march with more ease and comfort than I anticipated. The weather was cool and I stand the march quite well, but a dream I had night before last is yet much on my mind. I dreamed that my Mattie was then living, but circumstances required that I should leave her and marry another, and that I had contracted such marriage, with her knowledge and consent. We were separated for awhile, quite a long while, and that I had a pleasant time with my new spouse (to her). But by some strange circumstance, I met Mattie. In a little log house in a strange wild place my old love for her immediately (renewed) returned. I seized her, embraced her and wept. It awoke me and I was weeping. A little sadness continues to follow me.

Just at night it commenced raining and at near dark we were marched into a piece of woods to camp, while it rained dreadfully. Oh what a cheerless night, and yet the boys would sing Union songs, and sing them quite lustily, too. The night was passed with half rations for supper of fresh beef, and in keeping up the fire, with but very little rest or sleep. And now I think of it I will have to mention the amount of fare we have had during the march. Sunday, had the fight without any breakfast, were taken and marched 28 miles, stopping near 12 o'clock at night. Laid down in the rain till daylight, then marched about ten miles and got about 1 pound of fresh (nearly so) beef. About 11:00 Monday marched again till dark and got one pound of flour and half pound of beef

View of Mount Jackson, Va. (from "Frank Leslie's Illustrated Famous Leaders and Battle Scenes of the Civil War.")

had some with salt. Tuesday marched all day without rations till dark received then three crackers, three-fourth pound of beef with salt. Wednesday and Thursday same. Friday marched all day, had only one-half pound of beef for supper and laid in the rain all night.

October 24 - This is Saturday and the seventh day of our march. We are at Stanton[5] having marched seven miles through a light rain and camped in a shed stable partially cleaned. An abundance of hard bread, some poor bacon and plenty of fresh beef was furnished us.

October 25 - Sunday and just one week after our capture we find ourselves all ready to take the "cars" for Richmond and we were agreeably surprised to find ourselves on board well-warmed passenger coaches, with a very pleasant Pheonix[6] and a fine undulating country so far. Appears to be getting more level but mountain tops are to be seen in the distance in various directions. Waynesborough county thus far about the same. The village has perhaps five hundred souls. Greenwood is east of two tunnels in the blue ridge, west three-fourths of a mile, the other sixty rods. Of course it is very rough but a nice valley is to the right of us and below us. Charlottesville 3000. Nice and pleasant country east of Blue Ridge. Gordonsville three or four thousand or thereabouts, country getting undulating. Seems to be well defended by artillery. Have been sleeping and missed the country back of Charlottesville. My eyes are very sore and annoy me very much. Well, saw that much of the country is very

---

[5] Actually Staunton, Va.

[6] A thing of uncommon beauty. Phenix is the preferred 1860s spelling.

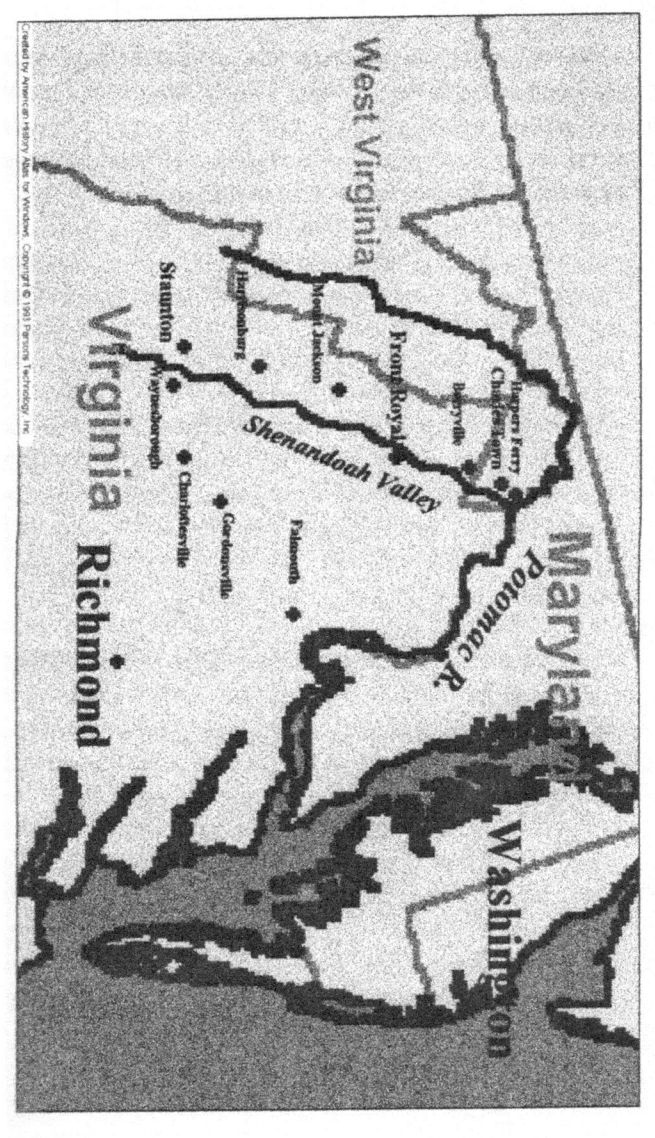

Created by American History Atlas for Windows 1.0, copyright 1993 By Parsons Technology Inc.. Used by Permission.

level, wet and uninteresting. From Gordonsville to Richmond there was nothing worth mentioning. Arrived at Richmond about three o'clock and were marched to an old tobacco house near Libby Prison where we were turned in without any rations all day. Slept upon a dirty floor without any covers or anything under us.

Pvt. John G. Anderson of the 6th Michigan Cavalry's Company I. He was absent at Charles Town, having been captured June 11, 1863, at Seneca, Md. Anderson returned to the regiment Nov. 1, 1863, and was captured again March 10, 1864. He did not survive the war. (John Sickles collection.)

Look and Listen, Wish and Wonder

Stench of excrement and urine from the corner privy permeated the air in the warehouse prison, mixing with the stale odor of unwashed bodies. Lice swarmed the wooden plank floor. There was no escaping the continual itching as the vermin found their way into the prisoners' clothing.

No privacy was possible, no view beyond plain brick walls, a wooden ceiling and an occasional glance through barred windows. The best the crowded captives could do was simply endure.

When they marched through Richmond, Osborn Coburn and his comrades passed near Libby Prison. This converted warehouse already was notorious in the North for its squalid conditions. But the Confederates reserved Libby for officers. So Osborn and the other enlisted men were separated from their company officers and filed instead into a nearby brick building that, like Libby, was a former tobacco storehouse. Several such buildings were clustered near the James River waterfront. One was the Smith Prison — likely Osborn's new home — described as "sixty feet long by

View of an example of a Richmond warehouse prison, Castle Thunder. (From "Military History of Ohio.")

forty wide, three and a half stories high, containing four floors, the upper one, or attic, being very low and directly under the roof."

Osborn – with blue eyes, of average height at just over 5 feet 7 inches – kept his sandy hair clipped short and wore a neatly trimmed mustache. He would find such grooming difficult to maintain in the foul prison warehouse.

But Osborn didn't really expect to remain a prisoner for long. Since early in the war most prisoners from both sides had been paroled or exchanged. A paroled man gave his word not to fight until properly exchanged for one of the captured enemy, but once exchanged a man could rejoin the ranks.

In recent months, however, this exchange system had begun to break down. One problem arose over the Union's use of black troops. Southerners, seeking to maintain slavery, argued over accepting black Union soldiers as regular prisoners of war. Due to such difficulties, in May 1863 Union General-in-Chief Henry Halleck suspended prisoner exchanges. But in certain instances, by special agreement, the practice continued. Hope of exchange helped motivate the POWs, including Osborn Coburn.

October 26, Monday – We are up with the sun having slept soundly from previous lack of sleep and fatigue for there was nothing else to repose. We find here a few troops from Burnside's[7], Grant's[8] and Rosecrans[9] army,

---

[7] Maj. Gen. Ambrose Burnside, commanding the Union Army of the Ohio, at the time occupying Knoxville, Tenn.

brought in lately. The news from the various seats of war are full as encouraging as we could expect through Rebel sources, probably far less so than our own papers would have us believe. At 8:00 o'clock rations are issued to us. Just half rations of good soft bread and good boiled beef. I ate part of mine only as I shall need the balance more in the morning. The time is passed principally in walking the floor, telling war scenes, etc. We "go to bed" early to secure a favorable position on the floor. Wrote a short letter to father and Eva to be sent by flag of truce boat which is here.

This letter was addressed to Osborn's father, John F, Coburn:

Father,

I am a prisoner here much jaded by the march but otherwise well, & in as good spirits as could be expected.

Will you be so kind as to write. To H.F. Thomas, Napoleon, Mich. C.C. Fuller Esq. Big Rapids Mich. & Miss Eva Aldrich, Olcott, N.Y. and tell them of it, also as many others as you wish. Be sure to write to Edwin – now don't delay these favors – please –

---

[8] Maj. Gen. Ulysses S. Grant, recently appointed to command the newly formed Military Division of the Mississippi.

[9] Maj. Gen. William S. Rosecrans, who until Oct. 20, 1863, had commanded the Union Army of the Cumberland at Chattanooga, Tenn.

Circumstances require brevity, so you will excuse it I'm sure.

Yours as ever, J.O. Coburn.

The following day he resumed his diary entries.

October 27 - The sun rose tolerably clear this morning and it would be a treat to be able to promenade the streets of Richmond, but prisoners have no such privileges. It is quite a novelty as I was never before in any manner under durance, and can truly say that no earthly record of any kind can show ought against me. My name does not appear even as the party to any suit of law. O how I hope this may continue to be true and that the Great Book will show as clear a record for me.

God smiles outside in the genial rays of the shining sun and everything seems bright. We are told that a general exchange of prisoners occurs on the 1st of November. I trust that He is thus smiling upon us and that we shall soon cease to be prisoners, though it be but little more pleasant to be in any part of the army service.

I am certainly thinking of home and friends. One picture only have I left. All the rest are captured or destroyed. My money has been taken away from me and only Eva is taken from my pocket at times to smile upon me and assure me that she has remained with me as my guardian angel, and that I certainly shall outlive the war to be happy with her. The day passed on relieved only by our attempted game of chess. Myself and opponent are both amateur players hence but little real interest in the game. We received our usual 1½ loaf of bread and small piece of meat. We are told the bread ration will be doubled to us tomorrow.

Oct. 28, Wednesday ⁻ Looking from the windows we behold a pretty autumn morning outside, which has a slight influence upon the feeling of the prisoners within but we want liberty. The hearts of the men are cheered up this morning by the report said to have come from a Rebel officer that we are to be immediately paroled and sent to Point Lookout until further arrangements could be made concerning us. This may be so, I hope it is but I have long since learned to give little credence to "camp rumors". Oh how I long for something to read merely to while away the tedious hours. But nothing is to be had, so we sit and look and listen, wish and wonder till we are tired. Yet there is little gloom and discontent manifest. The men generally keep up a cheerful appearance and make the best of a hard fate. Well I have written so much early in the morning. Let us see what occurs during the day.

The sun is setting gloriously and from the windows of our prison room the scene is a golden one. The rippling water of the ancient James glistening beneath the soft mellow rays of the setting sun, the gold tinted leaves of the Shintbery near and distant forest, the milky azure of the blue expanse above are all beautiful and so suggestive of the greatness and grandness of Nature's works. One feels like seeing but to love them, but God's works as all else is viewed to a better advantage where liberty and a "full stomach" can assist in appreciating them.

Our rations today are the same except that the quantity of meat was considerably diminished. Just as

"Old Sol" has laid himself down I close to await the dawn of tomorrow, taking twilight to look at and think of Eva and many other things ere I sleep. I have slept but such a sleep. I may as well describe the manner of "resting".

The floor is swept, when the men arrange themselves in rows on either side with their heads to the walls, lying "close order" in the manner usually called "spoon fashion". Two similar rows are formed with their heads to the center of the room. We are without blankets or overcoats except in a few instances. Hence the "close order" to assist in warming each other with the united warmth of the body. Long before morning our bones become very tired from contact with the hard floor, which with the cool nights makes the night to us very unpleasant.

Various rumors continue to circulate as to the disposition to be made with prisoners, but I can place little confidence in any of them and conjecture is useless.

October 29 - Thursday. Another pleasant morning outside. All hands seem more cheerful this morning. More so than one could have reason to expect or than is usual. Wonder if it is the forerunner to our liberation or habitual calm that preceeds the bursting of the gathering storm. It is rumored that our pickets have approached to within four or five miles, somewhere (don't believe it) that Meade[10] has driven Lee[11] to this

---

[10] Maj. Gen. George Gordon Meade, commanding the Union Army of the Potomac in Virginia.

[11] Gen. Robert E. Lee of the Confederate Army of Northern Virginia.

side of the Rappahannock (quite reasonable) and that he has orders to move on Richmond if possible. (Possible it is so but if true how did the report get here.) So no confidence is to be given to reports. It has been told us also that no exchanges or paroles are to take place during the war. The truth is, the man telling it though a Rebel officer knew no more about it than I did.

O such a time as all hands are having cleaning floors here. The whole building was very filthy and the lice were intolerable. It is to be hoped the cleaning will partially end them for a while. All kinds of rumors concerning exchanges. That is and will be the all absorbing topic so long as we remain here. Am trying to sell watch for something to eat. I did think I could not part with it but hunger will make me yield I guess. One has to be very cautious too, as the only means of effecting such a sale is contraband and it won't do to expose either myself or parties buying. I would sell almost anything I have for bread but that which was given me for keepsake as especially Eva's presents. Those I will not sell but to save life.

Well I declare it is 4:30 o'clock and nothing since 2 p.m. yesterday to eat and only half rations then. It is rather amusing to witness the effect it has on different persons to become so hungry. Some are cross. Some despondent. Some make light of it. While the greater share seem to be content pacing the floor and minding their own business, mostly thoughts. Well, how do I feel and act? I feel hungry. That's certain. Well I have been lying upon the "less table" doing nothing and seeing nearly nothing until it suddenly entered my mind to note down something about it and here it is.

Just at dark our rations came. Bread the same and meat rather more than restored so that we made quite a comfortable supper, while the prudent ones here had a handsome piece of bread for the morning. Like the swine in the sty, we eat and lie down. There is one benefit arising from this course. It prevents a rapid digestion.

30th Friday - As usual all hands are on the move with the dawn of day. Eat the little bread left from last night's ration and commenced talking over the prospects of release. Last night there was an arrival at this building of some 200 or more prisoners from Mobile who had once been paroled but something seems to have been wrong as all are yet detained.

Last night paper was brought into the room which states that all prisoners taken prior to Nov. 1 are to be paroled and sent to our camps of parole but not exchanged. All taken after Nov. 1 are to be retained and fed by the respective governments. That is the U.S. feeds the Union prisoners in rebel prisons and vice versa. Knowing the difficulties of them in the way of exchange the report looks highly probable. If it is so we may look for our return to Loyal regions within the next ten days, though no return to duty perhaps until the war is over.

Well I was much pleased to meet my old friend W.P. Montonye.[12] I cannot say I was surprised for I knew he was a prisoner somewhere and rather supposed he had

---

[12] William Perry Montonye of Big Rapids, Mich. A member of the 3rd Michigan Cavalry, Montonye was captured by the Confederates while on a late summer raid near Grenada, Miss.

been sent to Richmond. We are having a good visit. He is here with us I suppose we are all called eastern troops. This I trust will take us to Camp Chase, Columbus, Ohio, where I shall be near my friends should furloughs be granted to paroled prisoners. Rations are issued to us at 2 o'clock. They are as large as at any previous time.

Oct. 31 - The night before passed was much warmer than at any time since our arrival and I slept quite well, if we only had enough to eat. I had given up the idea of selling my watch on account of the reports alluded to but now again the prospect looks more dark and I have half a mind to sell it. I believe I will not take less than $50 for it though, until I have starved a while longer.

Near noon the clouds broke away and a warm sun was let in upon us, that is so far as raised windows and open doors would permit. Cheerful faces greet me on every hand. At least after rations are issued. Today I ate my entire day's rations at a single meal and was hungry when I had finished. Now I have nothing to eat, till we draw again tomorrow. I think I shall not do so any more. It is now getting quite dark and I'll close the book till tomorrow.

Let Hope Predominate

Nov. 1 - *'Tis a very pretty morning. Oh how I wish I had my liberty to go out and enjoy it. But when that time will come I don't know.*

Well last night I was out on the steps and asked the guard if he wanted to buy a watch. He courteously invited me down to let him see it by the light of the moon. I did so and walked out of the building for the first time. We did not trade but this morning I sold it to one of the prisoners for $50. I did hate to part with it but bread is worth more than "time" although it is said that "time is money". Well some of our boys have had money all the time and plenty to eat, persistently refusing to lend for fear they might sometime get out. Before I used a cent of mine I loaned two of the boys $5 each. I don't see how some people can be so selfish. I have often wished I had enough to make them all comfortable. I'm sure I'd do it.

Well now that I have money all hands are trying to borrow, among them some who would not accomodate when they might have done so. After all I see that I'm a little revengeful and generally sad. "I'll see 'em d----d first." I declare it is Sunday and no bread can be bought. So I'll save so much today. And I've sold my jet ring for $1.00 and let Orner[13] *have the money. This makes $51*

---

[13] Pvt. Levi Orner of the 6th Michigan Cavalry's Company I, captured with Osborn Coburn at Charles Town.

in all received and $11 loaned.

Like a fool I gave one of the guards $10 to buy some sweet potatoes saying "Couldn't make change." So I am out a $5 Confederate note. Never mind, it is only lent, for I am bound to steal and rob from secessionists whenever I can find them till I have made myself good for all I lose by being in the army, if it is possible. This resolution is newly formed for I had always entertained some compunctions as to such a course until I am brought to a realization of the fact that stealing at least from the enemy is part of the duties of the soldier. So Rebs stand from under. I'll never show you mercy hereafter.

There is some commotion in the building. Those on the lower floors are ordered above. I think more prisoners have arrived. This concentration of prisoners from all parts is indicitive of something. I hardly know what but hope it is to prepare for our being removed from close confinement. Do for God's sake take us out of this. It is horrible! Horrible! Having bought some bread I shall lie down with a full stomach feeling satisfied with the allowance. The first time for two weeks and a day. Bought a plug of tobacco for $1.00.

A good many are playing cards and checkers. This with all my irreverance for the Sabbath I will not allow myself to do. In fact I prefer not to violate the Sabbath more than circumstances seem to demand. The trade I made to day I consider justifiable, but gaming would not be. Reasons are apparent. It is now sundown and a beautiful evening. So let me lie down with thought of the sublimeness and lovliness of nature, sweet dreams of Eva and a happy future hoping for some good news on the morrow. I sent by Cutler[14] and bought $5 worth of bread, divided it among the men and is usually the case got the smallest portion myself. I could have bought the whole of it in the north for 75 cents.

2nd Monday - Beautiful Indian summer morning, our boys feeling quite well having enough to eat. I may as well count my $5 as lost. Well "it's only lent". At noon and the day seems to drag on slowly. I have tried a game or two of chess, but can get up but little interest. Have eaten a piece of bread (25 cents worth) and called it a dinner. I have loaned out so that 24 hours finds $50 reduced to $25. What will we do when that is gone, I think I could get along myself but some of the men are much heartier eaters than I am, and they will not be able to bear up under it.

The prospects seem good now for our remaining here some time. Perhaps during the war. O, God, I hope not. But if such is for the best I'll try and bear it patiently. Confound the Rebels, why will they persist in their damable course? If they are clearly wrong why does a just God permit them to continue evil doing? Will the advocates of the doctrines of special providential manifestations and visitations give even a plausible reason?

I think I never wished to be out of the army so much as now. There is not in all this house one with whom I can sit and converse in an appreciative conversation and I have no range of companionship. Grossness, vulgarity, coarse games and ribald jests and conversation are so

---

[14] Pvt. Luman A. Cutler, another member of Coburn's company.

common that they fail to shock the ear, but are to me very unpleasant. O what demoralization is consequent upon this war. Will it be possible to restrain the passions of vicious men thus for a while nearly unrestrained when they again return to citizenship? I hope the home influence will at once soften them down. The preponderance will be so great in numbers, that it may, if comparatively good, succeed.

About 8 o'clock we were ordered to pack up and commence marching. When we were halted it was upon the celebrated Belle Island. A part, about 2½ acres devoted to our camp. I'll leave a description of this horrible filthy place to some other time. Suffice it now that we were turned into its dirty streets to "Get along the best we could till morning", without overcoats, blankets or shelter of any kind and only one small stick of wood to ten men.

We shivered out the night with but little and some no sleep. I succeeded myself in crawling in a tent door and lay close to a man so that our latent heat permitted me to sleep three hours. After that I walked around till day.

A mile's march from the warehouse prison, west through Richmond's streets, south over a bridge spanning the James River and then west again brought Osborn Coburn to a narrow covered railroad bridge leading from the river's south shore to Belle Isle. Once across on the island, looking north at higher ground

Bridge from the James River's south shore to Belle Isle. (Massachusetts Commandery Military Order of the Loyal Legion and the U.S. Army Military History Institute.)

69

above the river's bank, Osborn could view the dim lights of downtown Richmond.

Tuesday's welcome, warming sunrise revealed a camp crowding the tip of Belle Isle, on a low flat projection at the island's north end. The soil was sandy, gray with a hint of orange, and polluted with the accumulated filth from thousands of prisoners. Small stones worn smooth by the river's current littered the ground. Marking the prison's borders was an embankment about three feet high and ditched on either side. A smaller mound inside the embankment indicated the "deadline," so named because any prisoner passing that point would be shot. Guards, bayonets fixed to their loaded muskets, paced back and forth outside the piled dirt border.

Belle Isle was about a mile long by a fifth of a mile wide, stretching upstream from the prison compound. South of the camp the island rose abruptly, a steep prominence jutting high above the river. Cannons stared down from the high ground, ready to rake the prison with shotgun-like blasts of canister should the captives attempt a breakout.

Shallow churning rapids of the James River surrounded the island. This current threatened to drown anyone who should elude the guards and attempt a swim to freedom. Downstream the prisoners could see the Long Bridge, a rickety railroad trestle supported on stone pillars. Clearly visible along the waterfront to the north were the red brick buildings of the famed Tredegar Iron Works, busy churning out cannon, ammunition and other material vital to the Rebel war

View of Belle Isle prison camp from the island's high ground overlooking the compound. Downtown Richmond, including the Confederate Capitol, is visible in the background. (From "Frank Leslie's Illustrated Famous Leaders and Battle Scenes of the Civil War.")

effort. Other structures, among them church spires and the white-columned Confederate Capitol building, dotted the higher ground behind the iron works.

Space remained on the island's low north end to expand the prison compound several times over, something the Rebels would find necessary to accommodate the growing number of POWs. Eventually as many as 10,000 captives would crowd the Belle Isle camp.

As another Michigander - John Ransom of the 9th Michigan Cavalry - observed when he arrived at Belle Isle two weeks later, "The River between Richmond and the island is probably a third or half a mile. The 'long bridge' is near the lower part of the island. It is a cold, bleak piece of ground and the winter winds have free sweep from up the river...

"The prison is in command of a Lieut. Bossieux,[15] a rather young and gallant looking sort of fellow. Is a born Southerner, talking so much like a negro that you would think he was one, if you could hear him talk and not see him. He has two rebel sergeants to act as his assistants, Sergt. Hight and Sergt. Marks. These two men are very cruel, as is also the Lieut. when angered.

"Outside the prison pen is a bake house, made of boards, the rebel tents for the accommodation of the officers and guard, and a hospital also of tent cloth. Running from the pen is a lane enclosed by high boards going to the water's edge. At night this is closed up

---

[15] Lt. Virginius Bossieux.

Belle Isle opposite Richmond, 1866 print published by C. Bohn, Washington. (Library of Congress Prints and Photographs Division.)

by a gate at the pen, and thrown open in the morning. About half of the six thousand prisoners here have tents while the rest sleep and live out of doors."

As another writer described Belle Isle, "At the very gate of the capital of the rebel government the prison lay, where men were penned, naked, mad, diseased, starving, living in filth, burrowing in sand like savages. The island contains about one hundred acres; the prison camp, established at the southern [actually northern] end, contained about ten acres...From the fall of 1863 through 1864, the ratio of mortality here, proportioned to the number of prisoners, was equal to that at Andersonville. The hospitals were unfit for stables, and patients seldom admitted until they were ready to die...With the James encircling three sides of the camp, the prisoners were not allowed to get water from the river, but were compelled to use it from a canal dug for the purpose, leading from the river to the camp, describing a semi-circle and returning to the river at a lower point. The canal was so arranged as to form a current through it, but the sinks were situated where their drainage found its way into the water as it entered from the river. The water could not be used for drinking purposes in the warm months, and the prisoners sunk barrels in the sand. The water obtained in these was better, but still tainted by the seepage from camp filth and deposits. Some A tents were issued, which provided no outlet for smoke, but these were greatly insufficient in number for the prisoners in the coldest weather."

Making life even more unpleasant was a gang of thieving Union men, termed the Raiders, who robbed and terrorized their fellow prisoners.

Thrust unprepared into the open-air camp, Osborn

continued to record his experiences.

Drew for ration three small hard crackers and one ounce of boiled beef each. Confederate money won't help here at any price so that I am without funds again. I was appointed sergeant of a squad this evening, and the charge and care of 100 men so long as we are here, devolves upon me. Went to draw rations for the men and while we were bringing it in the "raiders" stole four loaves. Being allowed but one quarter of a loaf to a ration it cut 16 men out of their bread. I divided my own ration and a small loaf I had brought with me so that five or six got a little.

3rd Tuesday ‾ It was a very warm and pleasant day. But O how crowded, and the stench around this miserable camp is horrible. However, I have got along very well. We had 1¼ loaf of bread and one ounce of meat in the morning, and 1¼ loaf of bread and one pint of soup at evening. I shall not "go to bed" hungry as I was fortunate in getting plenty of meat in our soup for all the men.

I took quite a hard cold last night. I fear it will rain tonight and if get no shelter it will be hard to endure.

Tents are here but they do not seem to be issuing them yet, and it is near night. I have been considerably annoyed by persons who are not assigned to messes. Chiefs of messes are not altogether honest and crowd out some who have been counted in their messes.

There are many sick, once I observed while walking through the camp. They are really in a distressing condition. Far away from home and friends, cut off from the assistance our benevolent government would be glad to give them. Pale, haggard, ragged, dirty and miserable in the extreme. The enemy lends them no more pity or assistance than they would a beast.

Poor, poor men, how I do pity them, many of them must die, aye, nearly all, and be buried with no pitying, sympathetic heart or tearful eye. They die and are buried unknown to their friends. A distant parent at home wonders "how my son is". The fond wife has not heard from her husband for a long time. Aye, and will not until years perhaps have passed, when by chance as it were, some stranger tells her he knew of his death in a prison camp. The maiden fair wonders in her heart where her lover is. She knows of the hard fought battle in which his regiment was and learns that he was among the missing, but nothing else is heard of him. He is not among the dead upon the field. No he was a prisoner. Carried to this accursed place and had passed away. And the maiden wonders on long years, pines away, droops and is as the faded flower. Her lover does not come.

O, Eva, Eva, will it be true with you? Must it be my lot to thus pass my remaining days, when I had pictured in my mind such a happy future with you? No O, no, God forbid that it should be thus. I will let hope predominate and feel determined that I shall live to escape and see the end of all this trouble. I will always be cheerful in spirits and careful in habits, and trust in a kind Providence. Eva, dear, send your prayers to Almighty God in my behalf, and also, for all our friends in Union army. I am inclined to a sadness of heart tonight. Dwelling as I am upon it but adds fuel to the fire.

Another incident and I will close tonight. Gen. Neal Dow was passing among the prisoners inquiring after their wants and says the government is going to assist them.

O, how the hearts of these patriot patient prisoners did well up and leap with that pride and joy only known to American hearts at the sight and voice of an American officer. I could not surpress the tears as I witnessed the good it seemed to do them, to see an American officer in uniform. They may be distressed, but they cannot be broken. In their true, loyal American pride and spirits God bless them and bless the noble heart in Gen. Dow for his interest in their welfare.

Without covering tonight though the weather is much more pleasant than last night. I am using up my book too fast as I may be kept here a long time.

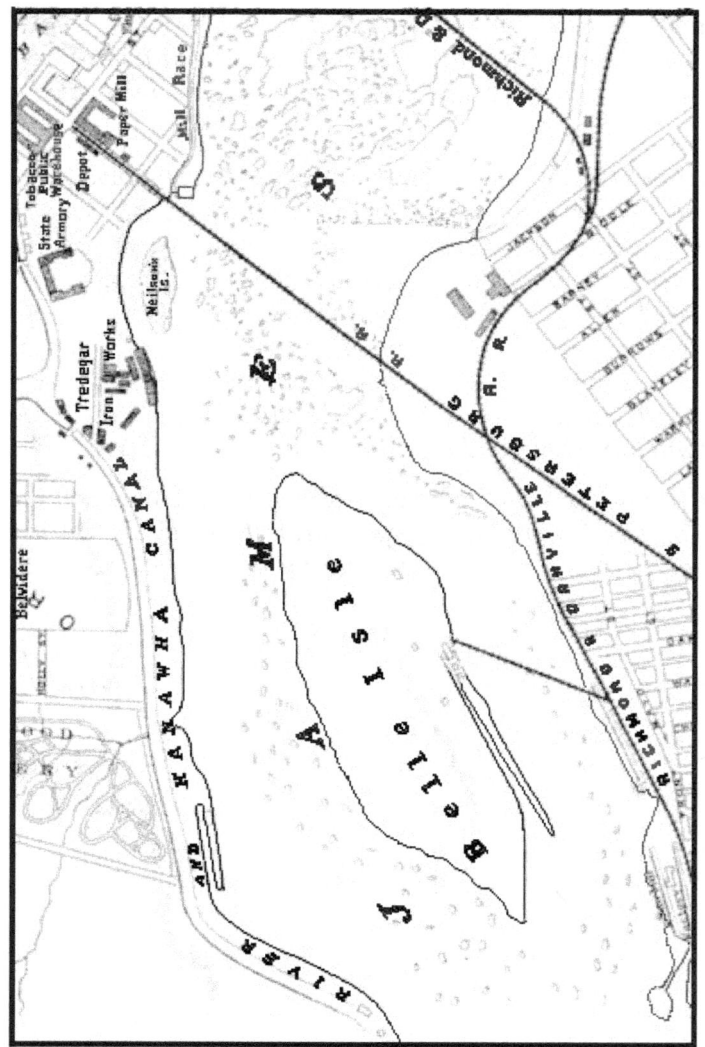

Belle Isle and the James River waterfront. The prison camp occupied the low northeast (right) end of the island. (Adapted from "Atlas to Accompany the Official Records of the Union and Confederate Armies.")

O, Won't We Be a Happy Crowd

Neal Dow seemed almost a savior to the Belle Isle captives, the venerable general with his long, snowy beard and white hair passing among the pressing masses. Prisoners saw in the general a welcome sign that someone, somewhere, cared about their welfare.

Before the Civil War Dow gained fame as a prohibitionist. Instrumental in passing an anti-liquor law in Maine, he also favored abolition of slavery. Although 57 when the war broke out he entered the service, gaining a colonel's commission and later promotion to brigadier general. After his capture in June 1863 near Port Hudson, La., the Confederates investigated Dow on charges of inciting slaves to leave their masters, and urging fugitive blacks to organize. However, the southern officials found insufficient evidence to try him.

Now incarcerated in Libby Prison, Dow turned his energies toward improving the lot of his fellow captives. By way of U.S. Christian Commission delegate John Hussey, Dow sent notice to Washington calling attention to the prisoners' plight and asking that supplies, food and money be sent to their aid.

"He declares," Hussey wrote Nov. 7 regarding his conversation with Dow in Richmond, "that the soldiers on Belle Isle are suffering beyond endurance. There are

5,400 on the island, which is low and unhealthy. They have not tents, into which by crowding more than one-half can enter at all; the remainder sleep without on the bare ground without sufficient clothing and almost entirely without blankets.

"Many have no pants; many have no shirts; so of shoes; and almost every individual lacks some essential article of clothing. They are on half rations, have no fuel of any kind, no soap is issued to them; they are consequently very filthy, of necessity. They need socks, additional supply of blankets and clothes, unless exchanged soon; shoes, mostly 8, 9 and 10.

"They are dying at the rate of eight and ten daily now, and the rate must fearfully increase from this on. One hundred will die daily by January 1. The general says they ought to be exchanged if possible, or many, many lives will be sacrificed and the health of the most of the remainder impaired."

Also bringing notice to the captives' suffering was Assistant Surgeon S.J. Radcliffe. The surgeon described the condition of 189 sick and wounded prisoners from Belle Isle sent north by ship under flag of truce. The ship arrived at City Point, Va., on Oct. 29. "Before she arrived at Fortress Monroe four died," Radcliffe wrote from Annapolis, Md. "On the trip from Fortress Monroe to this place four more died, leaving 181 to be admitted. To express fully the condition of this number language is almost inadequate, and none but those who saw them can have an appreciable idea of their condition.

"I do not pretend to particularize, for every case presented evidences of ill treatment. Every case wore upon it the visage of hunger, the expression of despair, and exhibited the ravages of some preying disease within, or the wreck of a once athletic frame. I only generalize them when I say their external appearance was wretched in the extreme.

"Many had no hats or shoes, but few had a whole garment, many were clothed merely with a tattered blouse or the remnant of a coat and a poor apology for a shirt. Some had no underclothing, and I believe none had a blanket. Their hair was disheveled, their beards long and matted with dirt, their skin blackened and caked with the most loathsome filth, and their bodies and clothing covered with vermin. Their frames were in the most cases all that was left in them.

"A majority had scarcely vitality sufficient to enable them to stand. Their dangling, long, attenuated arms and legs, sharp, pinched features, ghastly cadaveric countenances, deep sepulchral eyes and voices that could hardly be distinguished (some could not articulate) presented a picture which could not be looked upon without its drawing out the strongest emotion of pity.

"Upon those who had no wounds, as well as on the wounded, were large foul ulcers and sores, principally on their shoulders and hips, produced by lying on the hard ground, and those that were wounded had received no attention, their wounds being in a filthy, offensive condition, with dirty rags, such as they could procure, encrusted hard to them. One man who died on the trip from Fortress Monroe told the surgeons previous to his death his wounds had not been dressed since the battle of Gettysburg, Pa., where he was wounded in the head

Lithograph of a released Belle Isle prisoner at the U.S. General Hospital, Annapolis, Md. (From "Narrative of Privations and Sufferings of United States Officers and Soldiers While Prisoners of War", U.S. Sanitary Commission.)

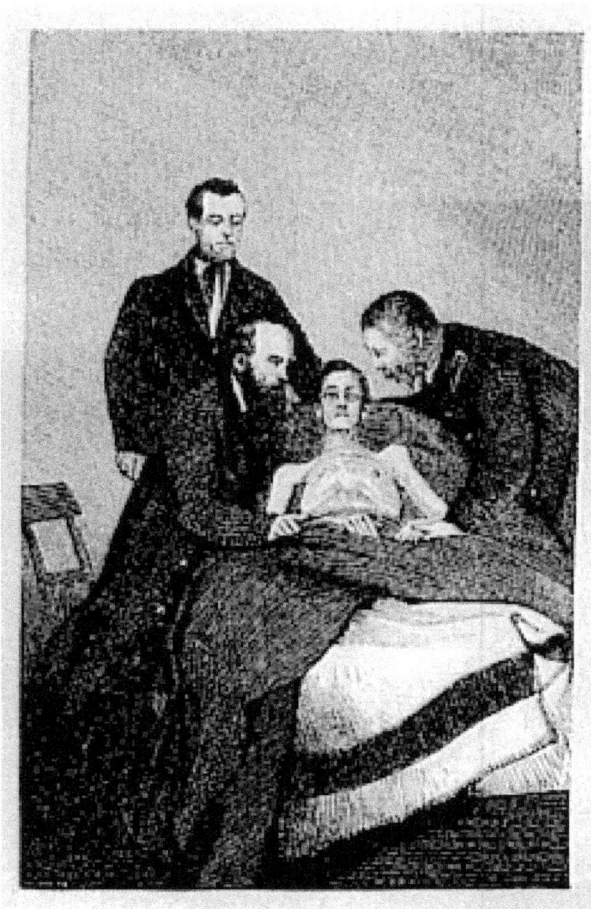

Lithograph of a released Belle Isle prisoner at the U.S. General Hospital, Annapolis, Md. (From "Narrative of Privations and Sufferings of United States Officers and Soldiers While Prisoners of War", U.S. Sanitary Commission.)

Lithograph of released Belle Isle prisoner Jackson Broshears of the 65th Indiana at the U.S. General Hospital, Annapolis, Md., in May 1864 nearly eight weeks after his release. (From "Narrative of Privations and Sufferings of United States Officers and Soldiers While Prisoners of War", U.S. Sanitary Commission.)

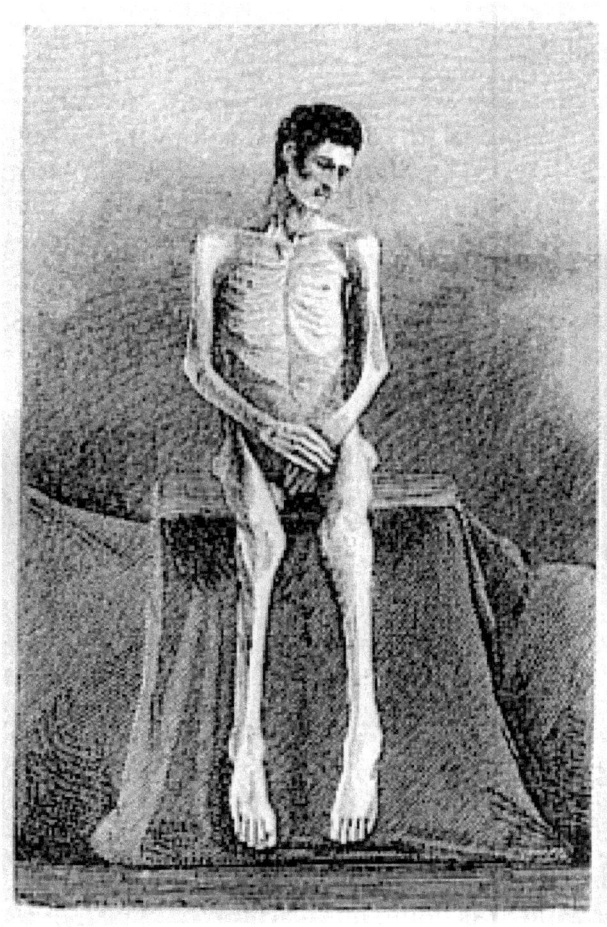

Lithograph of a released Belle Isle prisoner at the U.S. General Hospital, Annapolis, Md. (From "Narrative of Privations and Sufferings of United States Officers and Soldiers While Prisoners of War", U.S. Sanitary Commission.)

and both tables of the posterior part of the skull fractured.

"A majority of the cases were suffering with diarrhea, some of them with involuntary evacuations, their clothes being the only receptacle for them, and they too weak to remedy the difficulty. This being the case, you can of course imagine the stench emitted from them.

"Many had pneumonia in some form or stage; some were in the last stage, some gasping for their last breath. Delirious with fever, many knew not their destination or were not conscious of their arrival nearer home; or racked with pain, many cared not whither they went or considered whether life was dear or not. In some, life was slowly ebbing from mere exhaustion and the gradual wasting of the senses.

"How great must be the mortality, then, of these poor men, and how dreadful among those still suffering the pains of imprisonment..."

Such reports led the U.S. government to send rations and clothing the captives in Richmond, but the prisoners believed most of the articles ended up in the hands of the Rebels.

Osborn Coburn and the others on Belle Isle knew little of what was going on behind the scenes, but simply coped with their privations.

Nov. 4th Wednesday – The day opens bright and clear. I feel better from my cold and the camp disease[16]

---

[16] Soldiers' term for diarrhea.

which has been affecting me some has nearly disappeared. Bread gave out at No. 41, mine being 58 the last. Hope for a "reinforcement" soon, if none arrives we will probably get but one ration today. Only an hour passed till bread arrived and we had our usual ration. Commotion in camp because all taken in July and August were called out to answer to their names. It was rumored that they were to be paroled, but I believe no report but wait for the transaction.

I was much pleased with my old acquaintances from Ohio, Sergt. Wynn and Dan Will[17] from Williams County and Sol Miller[18]. Had a good visit with them. It is near ration time and I will close till tomorrow, or something occurs. The day closes and we lie in our camp street as heretofore secured a shelter for myself but lay upon the cold ground.

Nov. 5th, Thursday - Dark clouds floating high in thick broken scuds, indicates some wind and cooler days. Rain will follow ere long, and then what will we do. We are encouraged this morning by the report that several thousand are to be moved to Danville and Lynchburg. This will leave shelter for those who remain. Also that Grant has flanked Bragg[19] and taken Lookout

---

[17] 1st Sgt. Turner M. Wynn and Pvt. Daniel Will were members of the 100th Ohio Volunteer Infantry's Company C, captured at Limestone Station, Tenn., on Sept. 8, 1863. Both were residents of Bryan, Ohio, near Osborn Coburn's former home at Edgerton, Ohio.

[18] Sol Miller could not be positively identified.

[19] Gen. Braxton Bragg, commanding Confederate forces around Chattanooga, Tenn.

Mt. and that Charleston is about "gone up". These last are said to have been in the Richmond papers last night though I did not see it.

Surprised twice. In drawing rations meat gave out so that my squad was the only one without. I went to lunch of bread quite wrathy, but was soon sent for and got meat. One issue had been stolen, but was made up to me.

Gen. Dow is back today and a great many boxes, apparently clothing, is awaiting shipment across the water. I wonder if it is not for us. Weather breaking away and becoming warmer. I like to have forgotten to mention getting a double ration of bread for myself. I thanked the Q.M.[20] cordially and also for the pains he took to get us our meat. Yes the clothing was for us and was being issued to the men today. But much irregularity prevailed and all kinds of rascallity among our own men in getting more than was their due. Just at night it commenced raining very lightly and we feared we had a hard night before us, but were happily disappointed as little fell.

Nov. 6th, Friday⁻ The weather has broken away this morning somewhat and we hope for more pleasant weather than we anticipated last night but there are some ominous indications of a storm near. I hope all will get blankets and tents too. The day had nearly passed and the clothing all issued, only about one in ten were issued though a greater number of the actually needy.

---

[20] Abbreviation for quartermaster.

More are promised the first of the week.

O how tired I am getting of this lying around in the dirt with nothing to read. When will the commissioners come to some terms as to our release either by parole or exchange, however it serves to the interest of my country to remain I'll try to be contented but it is hard. The clothing that was issued I learn is furnished through the Christian Mission. We are thankful for their Christian kindness, but it looks like no extra rations from Uncle Sammies pantry. Well I can live yet I don't get enough to eat.

I suppose Gen. Meredith[21] arrives again today. I have no doubt but something will be accomplished but perhaps not this trip. Eva smiled at me today and I cannot discern the least trace of pity in her countenance. I shall certainly reproach her for her hardness of heart the first opportunity.

Tis near ration time and I close till morning. The night was clear and cold. We had five sticks of firewood which was all we had to protect us from the outdoor cold. No sleep all night. I was quite unwell with diarrhoea.

Nov. 7, Saturday - All hands feel poorly after so cold a night out of doors without blankets and no sleep. The loss has to be made up in part today. The camp ground is being enlarged and tents put up. This looks like shelter tonight, but no shelter comes. All the tents are being put up before occupying any of them, which will keep us out of doors until Monday probably.

---

[21] Brig. Gen. Sullivan Meredith was U.S. commissioner for the exchange of prisoners.

Major Thomas P. Turner, commandant of Belle Isle, dressed in gray with a felt hat, is the prominent figure in the foreground. Confederate guard tents are along the river in the distance. (Library of Congress Prints and Photographs Division.)

One more man dropped from my squad today, which makes two taken out sick to hospital and one arrested as a deserter from the Reb Army. Our own company keeps up remarkably well. This is fortunate. I lay it to their cleanliness and habits of diet. They are thus far very quiet and orderly and the best of feeling prevails among them, in fact, throughout the entire squad. There has been no quarreling and they look to me so kindly and seem to think I do all I can for them. I had not expected so much good feeling and feel very thankful for it.

I have not been well today, but think a good drink of red pepper tea this morning and a piece of camphor helped me. Got another extra large ration for myself tonight. I have missed but two or three times, but dividing it up with the others does me but little good.

Nov. 8, Sunday - The air is cold and raw though the night was warmer than we expected. Last night some 10 of the men escaped with two of the guards. One was killed and two more wounded. It is reported that Bragg has been thoroughly beaten. The Richmond papers do not speak of it in that manner. They are silent too about Charleston while it is rumored that it has surrendered.

There is apparently some excitement and movement in Rebel military circles. Soldiers of some kind were passing on cars but could not tell whether rebel troops or our prisoners. Hope we shall be able to learn today.

Bought $5 worth of tobacco and divided with the men. Was small enough to sell fish on ration (my extra) for ten cents and bought some "Richmond coffee" so that I made quite a breakfast at 12 o'clock noon on bread, beef and rye coffee. Corn bread was issued to 3100, wheat bread to 300 when all gave out for two or three hours.

I feel strong in the faith that we will not be kept prisoners more than two or three weeks more. I don't believe the rebel government can afford to keep us. They have ceased to enlarge the camp, refused letters to be sent away and many more substantial though not easily described reasons for thinking as I do. O, won't we be a happy crowd when once on board a transport for our lines. God grant the day will come soon.

Nov. 9, Monday - Last night was very cold and yet the authorities would not let the men occupy the tents that were already pitched. So some 1,800 men lay in the street with but a small fire. It was really shivering their life rapidly away. If God is just in punishing nations and men for their sins, surely this so called Southern Confederacy will have much yet to bear. I believe He is and that a terrible retribution is awaiting.

We have been prisoners now three weeks and a day, that is those captured when I was. We have endured much suffering from hunger and cold, yet there are many who have been here since the great battle of Gettysburg. They have endured longer, but little more than we. They had warm weather and most of the time tents and have the latter yet. They have also secured many blankets while we have none. The clothing issued by Gen. Dow the other day was also given them as they were near destitute so they are comparatively comfortable. But we are really suffering.

The night passed. High cold winds, a little rain and snow falling make it terrible for the poor fellows lying out of doors. To make it worse, all we had to eat the whole day was four ounces of corn bread. Double rations promised in the morning.

November 10, Tuesday – The clouds have passed away but the wind continues high, our men were many of them without wood as well as shelter. They are chilled through and in a suffering condition. My heart bleeds for them and would to God I could alleviate their sufferings. But I am in as bad condition as the rest with no power to do anything.

The morning paper today states that it is known that Meade has orders to hunt Lee and fight him. That he has already captured 3000 to 5000 prisoners. I sincerely hope he will be sincerely successful and that it will assist in our liberation. We are literally freezing and starving. Surely our country will not permit much longer. We must have something done or all shall perish in a little while. All but about 500 are now in tents, but without blankets or wood the cold is intense.

But why speak so discouragingly, when I feel so confident we shall not be detained but a few days. I still feel sure of this. So drive away sorrow, despair and despondency. Come in faith, hope, cheerfulness. Our sufferings are but for a few days. This being the darkest just before the joyful news of our liberation. "So let the wide world wag as it will I will try and be cheerful still." But the night draws nigh cold and cheerless. Rations of corn bread and boiled bacon are issued late.

O Dear What Shall We Do

Nov. 11, Wednesday - The morning is clear and cold and the suffering among the prisoners was intense. From 300 to 400 lay out of doors with no protection from the cold, and those in tents have to lie upon the ground which is damp and cold. My God is there no help for this. Does the best interest of our country really demand it, if so, God grant cheerful resignation upon our part.

We are pleased to learn that some were taken away early this morning. This number being variously estimated from 50 to 500. I think about 100 to 200 is correct. Well it looks as if something was being done though the commencement is small when there are 11,000 or 12,000 to be disposed of.

Rice and meat both gave out at supper and a large portion of corn bread was issued to those without. I got more for my squad than they could eat, thankful.

Nov. 12, Thursday - The morning clear and cool but more moderate than yesterday. It is stated in the Richmond papers that Raleigh, N.C. has been captured by our forces. This looks like [undecipherable] move but success in the movement of our forces. It does look like the Reb. Govt. must cave in soon.

Some 600 prisoners are brought to this miserable

camp this morning but no more shelter. O dear, O dear what shall we do. Two men were carried from camp dead. They were denied admittance to the hospital and so have perished. O Jeff Davis call on Almighty God for His assistance to sustain your rotten invention of a government. It is all a mockery and but bringing a greater curse upon you when the day of retribution shall have come.

I dreamed of Eva last night, it was a sweet and pleasant dream. I thought I made her a visit "in my suit of blue" and that we were finally married before we came back. I thought we were really happy. I awake and it was all a dream and I in this miserable camp. O how sad I did feel, but will live in hopes for the best. It often seems as if it would be such a relief if I could but write her and receive her sweet, interesting neat little love letters. But even that is denied me.

Our rations for the day were 8 ounces boiled rice for breakfast; at noon we got ration of bread. At night we got no sweet potatoes as did many of the squads but double of corn and rye bread.

Nov. 13, Friday - A very pleasant morning with a fair night's rest and a strong hope that we will not be kept here a great while. Quite a commotion in camp. All the sick were taken out. A similar move being made about a year ago when all others were marched out and afterward paroled. Have but little hope that such good luck will immediately attend us. Clothing issued to men today, about 150 sick taken out of camp. Very many for the time of year. Reported that Meade [undecipherable]

Belle Isle and its prison camp, with Richmond visible across the James River. (Reproduced from the Library of Congress LCUSZ62-12809.)

are at Orange C.H.[22] Ropes seem to be tightening upon the Reb. government. Have been lucky in rations for squad and self. Night drawing nigh. Sold tonight on rations. Only raw potatoes and bread.

Nov. 14, Saturday - Pleasant weather continues. Not a cloud to be seen but upon the countenances of the more discouraged here, and who can really blame them. Some have been here several months, and are getting very destitute of everything with no evidences of relief from their imprisonment soon. They are becoming very emaciated, many having died. Yesterday three were taken out of the camp dead. How many from among "are sick and in the hospital?" I have no means of knowing.

What is there to buoy up their flagging spirits? Nothing but patriotism and resolution. I declare when I look upon the sad scene around me I feel very like yielding to despair myself. But knowing this to be one of the surest ways to dissolution and destruction I rekindle the diminishing spark of resolution and reiterate my determination to outlive it all and go to my friends once more.

Very good dish of rice and corn bread this morning, making it into a quart of warm soup assisted me very much.

Indications of a storm are visible in the conditions of the atmosphere. Think it will be rain. I am getting quite weak from want of enough to eat and some exposure. Have half a mind to write to Eva and see if I cannot get an answer, but thinking we shall be removed soon to some other place of imprisonment or to our lines

---

[22] Orange Court House, Va.

ere long I will postpone it a while longer.

Heard that Meade had advanced to Orange C.H., captured Fredricksburg, 1800 prisoners, 16 guns and warping it to the Rebs generally. God be praised if it is true. Also reported today that there is a change of commissions on exchanged parole. I predict that this looks to an immediate understanding. Hopes revive though not yet deferred. Small rations of corn bread only for supper.

Nov. 15, Sunday - Commenced raining just at dark and rained all night. Many men could not get into tents and suffered much. This morning the heavens were overcast with clouds while it rains at times. The prospect is gloomy enough.

The Q.M. gave out a little wood to make fires for those who were in the rain to dry themselves by, but those in tents, no matter how worn and poor the tents were, got none. This so called Confederate government must be in very straited circumstances or devoid of all civilized intentions to treat prisoners of war in the manner we are treated. I am inclined to think the former the principal cause while the latter doubtless has its influence.

We got no breakfast until 1 o'clock then a ration of bread and two of meat. Uncle Sam has sent us 84,000 rations, plates, cups and spoons, so reported from a source to be credited. Weather breaks away early in the day and becomes quite pleasant with a cool drying breeze.

Clothing is yet being issued and done in a very neat and systemized method. It begins to look as if no arrangement had been made or is likely to be made for our release. However I shall continue to hope as usual till deliverance does really come. Corn bread alone for supper.

Nov. 16, Monday ⁻ Another fine day overhead and prospects of fair weather ahead. 'Tis really a blessing. This makes a month and a day of my captivity. When taken I expected to have been inside our lines, exchanged or paroled long ere this and my hopes were good. Now the day seems as distant as ever, even more so. Save that I do not think our government will permit so many of her brave men to remain in so distressing condition during the winter. I have now determined to think that so soon as the fall campaigns have ended then the attentions will be turned to those who have been unfortunate in falling into the hands of the enemy. Perhaps it were as well to make no such calculations, but the mind wants something to speculate upon and what more apropos than our deliverance. Cutler was taken away sick today.

Nov. 17, Tuesday ⁻ The night set in quite clear and warm but it commenced raining near midnight, after a brisk shower it becomes cold, damp and very cheerless. It will be a gloomy, dismal day on Belle Island.

Camp disease prevailed more than usual last night and much suffering was felt. Nearly all of the squads have got a share of blankets. Ours got 50 which makes them more comfortable. Toward night the weather became much more moderate. Our supper is late and very poor, corn meal and soup.

No I Shall Not Die Here

Belle Isle's prisoners were comfortably clothed, well fed, and their sanitary needs provided for, according to a Confederate inspector.

Individual cases of hardship, the officer concluded, were often the fault of the prisoners' own negligence or their cruelty to one another.

Confederate Major Isaac Carrington issued that report Nov. 18, 1863. The major had inspected the various holding facilities in Richmond and presented his findings to Brig. Gen. John Winder, commander of prisons in Richmond. Carrington stated there were 11,650 Union prisoners in the city, including more than 6,000 on Belle Isle.

"The men are comfortably clad," Carrington wrote. "I observed some few of the prisoners who were suffering for clothing. The supply of clothing and blankets sent for them from the United States is now being distributed by officers of the U.S. Army selected from the prisoners."

"The encampment at Belle Isle contains 6,300 prisoners, all privates and non-commissioned officers, who are quartered in tents. The tents are pitched on an island, upon a dry knoll, from which the surface water is easily drained. The contiguity of the river renders the police of the camp easy. There is an abundance of excellent water, afforded by eight wells within the encampment. The camp is thoroughly policed daily. I observed that some of the tents were dilapidated by weather and some injured by carelessness in building fires. A supply of tents has been sent to the island to supply these deficiencies...

"I learn that during the present quarters there have been issued full rations of all the articles mentioned in the abstract, excepting meat. Owing to the large number of prisoners suddenly consigned to their care without notice, the officers have not always been able to provide a full ration of meat. The deficiency has never existed but for a short time, and whenever it did exist it was remedied as far as possible by extra issue of other articles.

"The ration now being issued consists of one pound of bread, half pound of meat, half pound of potatoes, rice or beans, vinegar, soap, and salt, according to the regulations...

"I doubt not but what there are cases of individual hardship and suffering; they are unavoidable in the management of such a number of men; they may proceed from accident, from the abuse of authority of a subordinate officer, from the neglect of the prisoner himself, or from the cruelty of his fellow prisoners... Some among them are in a very filthy and disgusting condition, simply because they will not avail themselves of the opportunities for cleanliness accessible to them."

A vastly different account of conditions, however, was given by a Union chaplain recently returned to the Union lines from Richmond. "General Dow had visited Belle Isle and found there a large proportion of our privates are without tents, barracks, or any shelter, herded like cattle on the cold or wet sand, lacking blankets, clothing, and sufficient food. He thought that those not already dying of starvation were being rapidly reduced to such weakness and exhaustion as would unfit them for military service on their return to our lines...On the day I left the prison, Wednesday, [Nov.] 12th instant, the entire ration to the officers was a piece of coarse bread, measuring just 2 1/2 by 3 inches. The small allowance of meat was then cut off from the officers, as it had been for some days before from the privates...

"On my way to Richmond from Columbia a Confederate official said to a naval surgeon who was with me: 'It is a hard thing to say to you, but your men on Belle Isle are dying of starvation'..."

The U.S. Sanitary Commission, a northern organization that looked into the situation, described the corn bread the prisoners received as being "of the roughest and coarsest description. Portions of the cob and husk were often found ground in with the meal. The crust was so thick and hard that the prisoners called it iron-clad. To render the bread eatable they grated it, and made mush out of it, but the crust they could not grate." This report also called the bread "half-baked, full of cracks as if baked in the sun, musty in taste, containing whole grains of corn, fragments of cob, and pieces of husk..."

Balancing out the ration, the Commission reported, was "meat often tainted, suspiciously like mule-meat, and a mere mouthful at that; two or three spoonfuls of rotten beans; soup thin and briny, often with the worms floating on the surface. None of these were given together, and the whole ration was never one-half the quantity necessary for the support of a healthy man."

As Osborn Coburn suspected, much of the prisoners' hardship was due to the Confederacy's circumstances. The southern troops in the field often did without sufficient food, uniforms and protection from the cold. At one point, in April 1863, a riot had erupted in Richmond because some of the city's poorer residents were starving.

But many believed – including Osborn – that not all of the POWs' suffering was accidental.

A Confederate House of Representatives committee investigated prison conditions, but its findings determined northern accusations were unfounded. The U.S. Sanitary Commission concluded that "Tens of thousands of helpless men have been, and are now being, disabled and destroyed by a process as certain as poison, and as cruel as the torture of burning at the stake, because nearly as agonizing and more prolonged."

Negotiations on prisoner exchange continued but the sides could not agree on terms. Brig. Gen. Sullivan Meredith, U.S. commissioner of exchange, closed his latest letter proposing exchange terms with the following warning: "Confederate prisoners held by the U.S. authorities are at present well fed, clothed and housed. Should you decline the foregoing propositions I shall deem it my duty to urge upon my Government the necessity and the justice of rendering the condition of Confederate prisoners held by us as nearly as possible similar to that of ours held by the Confederates."

Exchange efforts remained stalemated, and the Federal government began making good on Meredith's threat. For Confederate prisoners in the north, rations and clothing allowances were cut and their ability to purchase additional articles restricted. At Johnson's Island on Lake Erie, Rebel POW's resorted to trapping and eating rats to supplement their reduced rations.

Osborn Coburn could merely speculate on the correspondence going on beyond Belle Isle's banks. He knew only what he could see, hear, smell and feel within the prison camp's enclosures.

Nov. 18, Wednesday - The weather is much more pleasant this morning, but O dear how sick I was last night with this terrible camp disease. I have not suffered more in one night since I came into the army. Vomiting and purging all night and nearly all day. In the morning I ate nothing but drank a cup of red pepper tea. About noon ate a small piece of ginger cake and drank a pint of rye coffee and begin to feel better. In the evening felt quite sick and persisted in eating nothing.

Intend to starve out the disease but what shall I do to rebuild upon. No bread but this miserable trash Indian meal, soup and sweet potatoes, all but the latter well calculated to make one worse instead of better. My God it is perfectly horrible yet I shall live through it, but am determined that it shall be my last service as an enlisted man. I will have a commission or a discharge. I have done enough, suffered enough and deserve some recompense besides the small monthly pay.

Nov. 19, Thursday - Fair weather overhead and could we be where Uncle Sam could care for us would feel quite contented. But here it is every man for himself, little chance to extend any sympathy to our suffering companions.

I feel much better in the morning and only need something fit to eat. As it is it seems almost impossible to revive from a fit of sickness. I must rely upon my own broken constitution and resolution and ever prevailing hopes. These are always better than medicine, even for disease. No I shall not die here. My predecessors all lived to a good old age to do good and bless mankind. I am not made to die younger than they.

Good news reported this morning. 7,000 to leave for our lines in a day or two. I don't quite believe it, but yet it might be so, hope it is, for heavens sake. Felt quite comfortable all day but must have worried around too much as I felt quite weak at night. Succeeded in trading my corn bread for some wheat biscuit. This with a cup of coffee and 1/4 of a miserable peach pie made my supper.

Nov. 20, Friday - The weather continued very favorable. But I am failing quite fast. My complaint is much better but strength is reduced.

This morning I succeeded in trading my rations for four small biscuits of wheat. I broke them into hot rye coffee and they tasted very good. They were better than those I had last night. I hope I shall continue to be as lucky, if so I shall get along, but it is enough to discourage the stoutest heart. Oh how I hope they will get an arrangement by which all can be sent to our lines. It is the one thing desirable here. But the interest of the government must be served though we sicken and die. Evening came very warm.

Nov. 21, Saturday - Rain commenced falling very lightly about sunrise. Oh I do wish for a better place than this if the winter storms are setting in. News of our exchange are rife, also that government rations have come for us. Their indications are antagonistic so whether we are to stay or leave is uncertain.

I feel somewhat better today but I am only able to just walk around a little. I am very weak. Hope that the worst is over and that I shall begin to gain, but I need a hearty diet which I cannot get here. I ate cornbread this morning the first for two days, having traded my rations for wheat biscuit, but all the cooking in the south is alike. I have never found but in one family a decent cook. They cannot make even passable bread of any kind. That lady was Mrs. Ott of Boliver who was raised and learned to cook in the north. Away with the sunny south. I have enough of it.

Nov. 22, Sunday - A disagreeable night was passed in camp last night. It rained all night, some of the time very hard. I was completely wet through in the morning, but notwithstanding felt quite well, having had quite a respectable breakfast of wheat cakes and toast of light wheat bread. It has cleared away in the morning and is really quite pleasant, but sickness and hunger is taking all the poetry out of me, though there was but little. Nothing to eat but corn bread with cold water to drink.

Nov. 23, Monday - It was clear and cold in the morning but before noon clouds overcast the sky and the air became colder. By two o'clock rain set in and the wretched starving soldiers were again suffering for the wants for the necessities of life. However they are more comfortable than formerly as nearly all are being quite well clothed and the balance will be as soon as they can be secured.

My squad was called out today and got everything in the line of clothing they actually needed but overcoats. I hope I shall be away from here before they are ready to issue them. But the prospect is rather discouraging although I cannot convince myself that we shall be compelled to stay here till spring.

The sickness in camp I noticed has somewhat abated. The cool weather I think is the principal cause. I am much better myself today. I think now that with proper care I shall regain my health. Government has actually sent us some rations. But the rascally rebel officers permitted their own soldiers on guard around us to be well filled with it before the hungry prisoners got a taste. Well, they were hungry too, for there is but little more issued to them than to us. The rest they have to buy. They have the chance to do so while we have not. Wonder if our government will retaliate as the affair was promptly reported to the officers issuing clothing and will surely reach our lines. Corn bread only for supper.

Nov. 24, Tuesday ⁻ A light but cold rain is falling this morning but the stomachs of the men are quite well filled with cornbread and beef sent by generous old Uncle Sam. This will counteract the disagreeable weather in part, but men ought always to have enough to satisfy hunger. My health is very much improved. So I think while lying still, but when I get up and try to exercise I find there is little strength left in me.

A lively report today that there is to be an exchange of prisoners. I'll believe it when we are moved from here to City Point, and slow to place full confidence in the report till then.

Our officers are not over today issuing clothing as usual. Wonder what that means. I don't think the light storm would deter them. This is flag of truce boat day. I just dare to think it might be so but not jubilant yet. Wish I could go to a comfortable fire in a comfortable position once more. I am very cold. My feet are suffering.

Nov. 25, Wednesday ⁻ Cold, raw and cloudy. Stormed a little just at night but before morning it cleared away. Was quite sick all day but drew the ration for the squad.

Nov. 26, Thursday ⁻ Thanksgiving day in all loyal parts of the country. My mind is forcibly carried back over the past, to contrast it with my Thanksgiving dinner on Peble Island.

Even but a year ago, while a soldier, I was hospitably entertained by Mrs. Poynton near Grand Rapids[23] *and a better filled table I never saw. I remember the lambs, turkeys, pigs, pastry, jellies and fruit of all kinds, devoured in years past. I remember the genuine prayers, praise and Thanksgiving, to Almighty God for His kindness, the gayety, mirthfulness, and almost universal happiness of the youthful. The generous donations to the poor, aye everything customary in our Christian, loyal America. But these things seem to have all departed from me as if by some magic influence.*

We had dinner here to subsist upon, the usual allowance of poor cornbread and meat, unless one is

---

[23]Grand Rapids, Mich., initial rendezvous for the 6th Michigan Cavalry. The regiment remained there until December 1862.

lucky enough to trade his rations for a cake or pie which may be found for sale upon the camp streets. I wonder if our president does really think of us in our misfortune here today and if he is not thinking of some honorable mode to accomplish our release. I believe him to be generous, kind and not unmindful of the wants of the people, and if that he really is thinking of Peble Island. O people of America let now your prayers ascend to Almighty God for the end of this cruel war and our deliverance from this miserable captivity. Join ye, in one accord let all disloyalty and divisions be banished to secure one great desideratum.

I see no immediate prospect of release. I fear we are to remain here till the end of the war. If so, God have mercy upon us, for it is the policy of these rebels to let us so far famish that our hearts and our spirits are so far broken, that if ever, it will be a long time before we are or can be fit for duty in the field, before relief is granted.

Well with such prospects ahead and with this book so nearly full, I must economize still more, barely making an entry of dates and leading events and leave it to be filled up whenever it shall please God and my country to grant me more ample time and means.

To Build Castles in the Air

Nov. 27, Friday - Rain and rain, cold day, everyone consequently cross and ugly. Squad becomes dissatisfied about my management and I told them to get someone else to act as squad sergeant. I feel sorry for the half famished men, yet it does seem as if they might use a little more reason than many of them do.

Well, the time will come though it may be distant when we shall be away from here and then these things will be mostly forgotten. But those who have borrowed money of me until I loaned myself out and then have abused me I never will forget. Our rations were short tonight.

Nov. 28, Saturday - Disagreeable rainy day. Short rations of bread and beef in a.m. Relieving myself from duty. Ate what little I got after that miserable cuss of a Payne[24] cheated me out of a portion, and then laid around in the tent and tending to my camp disease.

Rebel troops seem to be coming from the direction of Petersburg, reported that Grant has badly beaten Bragg[25] and conflicting rumors concerning Burnside[26],

---

[24] Pvt. Robert H. Payne of the 6th Michigan Cavalry's Company I.

[25] At Chattanooga, Tenn., Maj. Gen. Ulysses Grant's Western Federal Army routed Braxton Bragg's Confederate force in fighting Nov. 23 to 25.

that he is captured with 7000 men and that he is OK, near Petersburg. That a great battle must be fought between Meade and Lee. O God give us victory and then release us from this miserable place. Hope yet to be out, yet no evidence that we shall.

Nov. 29, Sunday ⁻ Rain, cold, wind and decidedly the most unpleasant day passed on the island. I am in poor health and yet was compelled to stand out in the cold storm while we were rearranged into squads again. I took a long jump from the 53rd to the 9th.

Segt. Howard of the 20th N.Y.S.M. who is decidedly a gentleman after my own heart in main respects, has agreed to try with me to get out of this. We are both sick, completely used up and should go out while we can. I do hope fortune will favor us once again. I think him in earnest in seeking a new home in the west, and he has agreed to visit Big Rapids positively. It is natural to build castles in the air, and we have anticipated some fine times when this cruel war is over and we are located as citizens of Mecosta county, Mich.

At age 44 Asa C. Howard was, like Osborn Coburn, older than most of his comrades in arms. A Mason before the war, the blue-eyed, dark-haired, light-complected Sgt. Howard stood 5 feet 10½ inches. When he departed his home in New York State to join the service he left behind his wife, Mary, and a granddaughter in their care, Mary White. Sgt. Howard's

---

[26] Maj. Gen. Ambrose Burnside's Federal troops at Knoxville, Tenn., were then under siege by a Confederate force under Lt. Gen. James Longstreet.

regiment, the 20th New York State Militia, was designated the 80th New York Volunteer Infantry when mustered into the federal service. But the men persisted in retaining their state militia designation.

The 80th New York-20th New York State Militia was in the First Corps of the Army of the Potomac. Sgt. Howard's regiment was engaged in the first day's fight at Gettysburg on July 1, 1863, and as Union troops retreated from the hills outside of town Asa was captured. He was far from alone, as approximately 5,400 Union men were taken prisoner at Gettysburg.

A first sergeant at Gettysburg, Asa was reduced to the ranks on July 31, 1863 by order of his regiment's colonel. No reason for the reduction was cited in Asa's records, although it may have been connected with his erroneous listing as a paroled prisoner. At Belle Isle Asa wasn't even aware of the demotion. But his wife, responsible for the household bills, surely was affected by the difference in income, at $13 per month for a private vs. the $20 received by a first sergeant.

Whether sergeant or private, Asa was someone with whom Osborn could relate.

Nov. 30, Monday - A little rain and very raw cold day. No wood and nothing for supper but the usual two ounces of meat. It does almost seem as if this infernal Confederate government desires the reduction of our numbers and was accomplishing it in this slow and barbarous manner of murdering us. I know they are hard pressed by our armies on all sides and their means cut very close, but they might certainly furnish us with wood to warm us and our rations.

O why are we kept here suffering so. We had to lie in our miserable tents rolled in our blankets to keep from freezing. Heavy firing was heard all day in the direction of Culpepper until late in the evening. It is said a great battle is coming off between Lee and Meade. I am hopeful as to the result and shall trust in Providence.

Dec. 1, Tuesday ⁻ Six weeks and two days a prisoner and four weeks on this island. I was in medium health when captured. Just recovered from my hurt in the June fight and the camp disease which annoyed me all summer. The march, fatigue, starvation, and miserable food brought the diarrhoea back upon me. It has been bad ever since.

I am now just able to walk around camp a little. Can scarcely lift my feet over an ordinary stick of wood. The weather is cold though clear today and I cannot exercise enough to warm my blood so I'll go to my tent, roll in my blanket to keep warm and think of the good things I used to eat, the happy home I left behind, the bright prospects I have so often conjured up I was to enjoy in another happy home with Eva, who looks at me, smiling on me as ever with no pity in her countenance, while hunger is gnawing at my vitals, and cold is pinching me. Nothing to eat but corn bread for supper.

Dec. 2, Wednesday ⁻ Weather more fair and comfortable, rousted out of our nests and had to stand in the cold till we were all counted in at two gates. Wonder if it is to find out the number to exchange. Guess not. Nothing to eat in the a.m. 250 more prisoners brought in last night. Poor fellows had no blankets and no tents. Fighting going on near Gordonsville, so reported. Sent to camp deans for rations.

Dec. 3, Thursday - Was quite warm and had I not been very unwell last night and today, should have thought it was very pretty weather. I became almost discouraged on account of the again declining health. In the p.m. it was reported the guards were to have eight days rations tomorrow to take us to City Point. Don't believe a word of it.

Dec. 4, Friday - Opens fair and warm for the season of the year. All hands out again to be recounted and resquadded. Presume the Q.M. had lost a loaf of corn bread. It's worth looking after in this d——d bogus government. Rested well last night and feel better this morning than for a long time which is quite encouraging. Got a loaf of wheat bread to eat before ration time which helped keep me up.

Am becoming very much attached to Sergt. Howard who sleeps with me, and we are fast becoming boon companions even in our trials and tribulations. He is sure earnest about going to Big Rapids to live. I sincerely hope he will as I am convinced he will make just such a valuable citizen as is wanted in that —— lot. O heaven grant that these bright anticipations may be realized.

View of Belle Isle from Richmond, with Tredegar Iron Works in the foreground. The prison camp was on the low ground at the left end of the island. (Massachusetts Commandery Military Order of the Loyal Legion and the U.S. Army Military History Institute.)

It Does Require a Stout Heart

Dec. 5, Saturday – *Day opens cold, clear and beautiful for the season. I am better in body and spirits than for a long time and ate a good breakfast of bread, butter and molasses, bologna sausage and Washington pie which made me feel still better. Good reports that a portion of us are to be exchanged. I do not think it reliable although it was announced in the Richmond Whig.*

Soap was issued to us today. I guess we all wanted to clean up, well it is time. Irish potatoes were given to some of the squads. Really the Confederacy is becoming extravagent to us Yankee prisoners. But towards night I began to feel worse and passed a very miserable night. The weather changing somewhat also and becoming colder made it very unpleasant to be up as much as I was compelled to.

Dec. 6, Sunday – *Clear and pleasant but quite cold. Was sick all day, eating only a small piece of meat, bread and molasses. O how terrible to live in such a place as this. My God is there no help. Kept my bunk all day.*

Dec. 7, Monday − *Clearer and colder than yesterday and I was much sicker. Answered to the doctor's call and got some powders. It checked my diarrhoea. At night I took another which corrected me in a short time. Was stupid all night from the medicine and was not up all during the night but suffered from the cold.*

Dec. 8, Tuesday − *Beautiful morning and the weather much more moderate. I feel much better and my spirits are also revived. Am prepared to eat and did eat a good healthy breakfast of sweet potatoes and a cup of U.S. coffee but felt worse toward night.*

Dec. 9, Wednesday − *Was not near so well today and had to lie in my blankets all day. Ate nothing for breakfast nor till night when I ate one hard tack, having drawn just half a ration in addition to our usual ration of corn bread and sweet potatoes. This hard bread evidently from our lines and being an extra I conclude is sent by our government. Rested quite well during the night, feeling better by noon, hope yet to get along well.*

Dec. 10, Thursday − *Feel much better again but shall worry round as little as possible. Soaked their hard tack in cold water till soft, then poured in more water, salted well and steamed well. I like them very well. The soup thus made was quite good at least so it seemed to me. Good reports are in circulation about exchanges. I shall permit myself to hope a little although my former hopeful disposition had given way to a conviction that there was no probable prospect of our immediate release. I am not discouraged but what I shall see the end, though I am quite low. Writing this as I lie on my miserable bed. I took a thorough wash on Tuesday in the open and I attribute the exposure (not* the washing) as part of the cause of my feeling somewhat worse.

Dec. 11, Friday ⁻ Weather rain and rather uncomfortable and I am not feeling better, after eating quite a breakfast of stewed crackers and crust coffee, conclude to keep as quiet as possible. But oh dear, it does require a stout heart to keep from being utterly discouraged, and yet if I allow myself to indulge in such feeling I shall assuredly go down.

I do wonder if it is the final conclusion of our country's officers to let us remain till the contest is decided by force of arms. If so our release may be near at hand but it appears far more distant than our constitutions will bear up under the treatment we receive here. I do not intend to think hard of my country or her rulers but this suffering is very trying to the soul. Oh Lord, remember those here that fought so bravely for a noble cause and do not forsake them in this time of need.

I must not fail here to mention the kindness I have received at the hands of Danny Will from Bryan, Ohio, he has done me several favors. Today he brought me a sweet ruck roll well buttered. O my but it did taste good. Large numbers of express boxes have arrived for the prisoners today.

Dec. 12, Saturday ⁻ Weather rather more favorable. But I was really worse today. Went out to the doctor and got some medicine. Ate nothing all day but a cracker and a half with two spoons of sugar. My physic worked off during the evening and night and I hope not to pass another day as sick and unpleasantly as I did this.

Dec. 13, Sunday - *Rained about half the night but it was warm and I did rest well after so sick a day yesterday. My appetite is somewhat demanding and I must have a plate of toast so here goes for one, such as I had to make it in this camp.*

Dec. 14, Monday - *Did not rest quite so well last night. This morning the weather is colder and somewhat invigorating. I go to get some more medicine and a cup of good beef tea[27]. Felt better when I find the last a little too much and could eat no supper.*

Dec. 15, Tuesday - *Was trotted out several times during the night and got wonderful hungry before morning. Forged an order, a German takes it and gets my beef tea, any port in a storm like this and must not be too honest. Decidedly unpleasant winter day.*

Dec. 16, Wednesday - *Was so much better that I felt highly encouraged. Appetite much improved, beef tea works well. Have plenty of it, had a good supper of fried meat and fried bread. Was somewhat disappointed in not hearing from Lt. Moore[28] to whom I sent for a little money. Begin to look for boxes from the company. Storm approaching.*

Dec. 17, Thursday - *Raining this morning but I feel*

---

[27] Beef tea is made by boiling a steak in water until all the juices are extracted from the beef. The broth thus made, lightly seasoned, was considered good nourishment for most cases of sickness.

[28] 2nd Lt. Malcom M. Moore of the 6th Michigan Cavalry's Company I, who was captured at Charles Town and remained a captive of the Confederates.

quite smart, much better than I had dared to hope for. My rations are getting too small again, I must get in extra some way.

Dec. 18, Friday - Felt remarkably well all day, but could not get enough to eat. Beef tea game having played out. They take it all down to the hospital and issue it there. The officers are issuing clothing and yet I get no word from Lt. Moore. Perhaps it has been forgotten. I am almost out of tobacco and don't know what I shall do. Cannot get enough to eat let alone trading rations for such things.

Dec. 19, Saturday - Came out this morning to see a nice clear but rather cold morning. Took a good wash and got ready for a breakfast but when ration call was sounded it was announced that no rations could be issued until 3:00. Cool I declare when a fellow is so cussed hungry. Well I am really becoming disgusted at the manner in which we are made to suffer here. No wood and nothing to eat. Well perhaps there is an excuse. The river is very high and it is perilous to get anything across but they have no business keeping us in so inaccessible a place, some secret monitor tells me we shall be kept here but a short time longer. Wonder if it is only a delusion. There must be an end to it sometime, why not soon. I'll allow a hope to revive me.

Dec. 21, Monday - Lively reports of Lincoln's Proclamation of Amnesty. The armistice of thirty days and our immediate exchange. The camp was wild with excitement towards night and I almost believed them on account of the tolling of the city bells which we heard.

Dec. 23, Wednesday - Some flakes of snow falling,

the first I have seen, but we get no breakfast and no apparent reason why. It does beat the devil that we must fare so damnably.

Dec. 24, Thursday ⁻ I am gaining now quite readily and begin to hope that I shall gain strength enough to do something to make extra rations. Have been hoping I could have something extra for dinner tomorrow but see no light.

Christmas ⁻ And nothing but the usual rations, and as hungry as a poor dog. Well it is the poorest and most cheerless Christmas that I ever experienced and that I ever expect to. The weather which has been very cold has moderated somewhat and we hope for better times even on Belle Isle. All sorts of rumors continued to circulate relative to paroles and exchanges but the cautious mind had schooled itself to become incredulous. So I believe nothing and wait for events and facts.

I Shall Yet Come Out All Right

Dec. 27, Sunday – *Was really the dawning of a better day although the weather was rather lowry and unpromising, yet the hearts of the prisoners were much cheered when 600 men were called out and left us under the announcement that they were to be exchanged, and that all were to be as fast as transportation could be procured. Yesterday there was nothing occurred more than the usual camp routine. Today all was excitement over the good news. Well this is rather conclusive to me that some understanding has been arrived, as there was no selection of prisoners the arrangement is a general one. O glory, glory! I do now hope in good earnest to see the end of my captivity soon.*

Dec. 28, Monday – *The day passes away and we are still here. It begins to look rather discouraging again, especially to me as my health is again growing poorer instead of better but yet I'll continue to hope on.*

Dec. 29, Tuesday – *Nothing of importance worth chronicaling.*

Dec. 30, Wednesday – *Will record the same as yesterday only that the day was the most pleasant of any for a long time. It was quite warm indeed.*

Dec. 31, Thursday ⁻ Was a marked change from yesterday. A cold rainstorm lasting all day and had but a small allowance of wood. During the night the water ran in so that it was an inch or two deep where I lay. I have wished many times since I have been here that I had quarters as comfortable as my father's hogs but actually I have not.

New Years 1864 ⁻ And we are yet here upon this accursed island. O how many anxious ones here had hopes to have spent this day with their friends at home or at least with friends in our own lines, instead of which we are dragging out a miserable existence as prisoners of war, and I doubt if any prisoners were ever before treated by any people professing to be a civilized nation as are the prisoners here. Thrown into old ragged tents without any bedding, blankets or anything of the kind. Our allowance of food is about a fourth what our government allows us. It is enough so that a hearty man will not actually starve, but not enough to prevent the rapid contracting of disease.

Why, O, why is this permitted. I have often been exhorted to place my entire confidence and trust in Providence in all kinds of emergency. That He would be our "stay and our support", would care for, protect and provide. Now we are here. The enemies cannon is in position to rake us at the first indication of a mutiny. There is no opportunity to assist ourselves. Can Providence? Will He do anything to relieve us, by our lending Him our blind submissive confidence? Well there is yet considerable hope that we shall be relieved soon. So I will not commence to murmur against God and country yet.

As the frigid air of January settled over Belle Isle the prisoners experienced a new horror, frostbite. If the poor rations didn't destroy a prisoner, it seemed the frigid chill would. Men huddled together, sometimes in holes dug in the ground, to stave off the freezing cold. Hands and feet of some were amputated from frostbite. By some accounts, the frozen feet of other prisoners were neglected until they rotted and dropped from the limbs, the death of the victim usually following.

"I have heard the men running around the tents to keep warm at all hours of the night," stated prisoner Walter S. Smith, a 48th New York Infantry private, "the river was frozen a little while I was there; the current is rapid. The water would freeze two or three inches in the bucket at night..."

"During January," testified Pvt. George Dingman of the 27th Michigan Infantry, "the men would run all night to keep warm, and in the morning I would see men lying dead; from three to six or seven; they were frozen; this was nearly every morning I was there; the men would run to keep warm, and then lie down and freeze to death; we made an estimate and found that seventeen men died a night from starvation and cold, on an average.

"If I were to sit here a week I couldn't tell you half our suffering."

This depiction of Belle Isle prisoners appeared in Harper's Weekly on Dec. 5, 1863. (Virginia Commonwealth University Libraries.)

Jan. 2, Saturday - Nothing new only the weather is now fair but cold and we have no wood. Yesterday's labor stewing bones brought me all I wanted to eat in bread and flour gravy. I wanted yesterday as a New Year's dinner but malicious ill health for a few days caused it to be one day late, but better late than never.

Sunday, Jan. 3 - Had an extra ration of cornbread today and am gaining strength rapidly. O, if I could only get enough to eat I would get right up I know full well, but if relief does not come soon I fear I shall certainly sink as have so many before me who were probably as rugged as I was.

Monday, Jan. 4 - The weather is growing colder and wood seems to be failing us. Our large supply in the tent is gone and they do not issue us more. I have frozen my feet already lying in bed and if the weather gets much colder I fear we shall nearly freeze to death or suffer worse than death.

Tuesday, Jan. 5 - The weather still growing colder. My feet so sore that I laid in the bed all day. O, My God, is there no help.

Wednesday, Jan. 6 - Same as yesterday.

Thursday, Jan. 7 - We got a small ration of bread and meat for breakfast. Ration of bread, meat and butter crackers for supper. O them butter crackers, how good they did taste. I never ate anything that relished so well or so it seemed. Tonight there was quite a fall of snow, say three inches. Of course the weather is a little more moderate on us prisoners, no wood and freezing.

Friday, Jan. 8 - After the storm it is clear and cold. The old 58th squad was called out for clothing and I had to stand weak and emaciated as I am and with frozen feet from 11:00 till nearly 2 o'clock till all had drawn. I did not get an overcoat as I expected but got a new shirt. I put it on and slept between it and for a long time, the night being the coldest yet. I suffered with my feet till 11 o'clock when I raised up and chafed my feet until they became warm and comfortable.

Saturday, Jan. 9 - It has been reported that no boat from our lines has arrived for two weeks. I think it is so for no mail has been received in this camp. This morning our mail has arrived and some express boxes which indicate that it has arrived, but brought in transports with Reb. prisoners dampened our prospects for a few days more. I do believe that every possible obstacle is thrown in the way of our release, but by whom is it done? Surely our government which has always cared with so generous a hand would not willingly let us suffer so much without an effort to relieve us. I wonder who the new commission to arrange a parole or effect an exchange is to be. It has been reported that Butler[29] was appointed but Secesh representatives refused to hold an intercourse with him. Will our government slide him off and appoint another or quarrel a long time over it? If a quarrel and we are kept here in consequence I shall certainly disown them. Well if the boat is really here we shall hear something

---

[29] Union Maj. Gen. Benjamin Butler, whose actions as military governor of Federally occupied New Orleans made him very unpopular in the South.

with regards to it soon, though how reliable we are always at a loss to know.

Sunday, Jan. 10, 1864 ⁻ It proves that no boat has as yet arrived, at least we have it and everything goes to indicate that such is the case. At all events we are here and no signs of being any where else soon. My feet are getting better and so is my health generally, and I manage to get something extra nearly every day. Weather quite cold.

Monday, Jan. 11 ⁻ Nothing has occurred of particular importance today so we will let it pass with this brief notice.

Tuesday, Jan. 12 ⁻ Will pass this day the same as yesterday.

Wednesday, Jan. 13 ⁻ Time passes on the same as usual. It will not pay to mention every camp rumor afloat, so with by mentioning the sale of beef tea for several days past to raise the extra and some slight hopes that we may yet get away from here soon. Butler has probably been relieved and it is rumored that Burnside will succeed him.

Thursday, Jan. 14 ⁻ Up in the morning early at work boiling bones. Sold 25 cents worth and had all I wanted for breakfast. Diarrhoea coming back on me again so I guess I shall require less ration today and for a few days to come than I have been even. News at night are rather encouraging. Butler instead of being superceded has been recognized by the Rebel government as the commission on exchange and that we are to be paroled immediately, which means in a few weeks. This news comes late in the evening and for the present as a camp rumor. Nothing more.

Friday, Jan. 15 - As on yesterday was up in the morning as soon as it was light but could not sell anything, luck seeming to have turned against me as I had been having remarkable good luck peddling heretofore. So I carried my merchandise back to my quarters and retired from business for the day. The news of yesterday is revived and considerable confidence seems to be placed in it. I have but little.

Saturday, January 16 - Feel quite poorly again today, was troubled a good deal last night but was somewhat better during the day, than during last night. O dear, the news is discouraging again today although it seems that a boat has arrived at City Point from Baltimore. But hope shall not fail, let reports be what they may.

Sunday, Jan. 17 - Last night got a little wheat flour, stirred it into my rice soup stewing it well, then sliced in my coarse bread and had a good hearty supper and one that checked my diarrhoea very much. This morning we got a good rice soup and made a good breakfast, but we don't get enough at night. We get nothing but a ration of bread and we all go to bed hungry, very hungry, and the prospect of release no nearer apparently than it has been. Shaved my face clean today, the first time since I was a soldier.

Monday, Jan. 18 - Lying in bed until after 9 o'clock in the morning waiting for ——— call. I commenced writing to pass away the time. It is three months since I was captured. Then I expected that all enlisted men would be paroled and exchanged and returned to our lines. We were full of health, heart, hope and spirits. We were fleshy, having known but little of hunger. We were confident in our ability to endure almost anything. Now we are down, clear down, starved out. Our flesh as well as hope and spirits are all broken or nearly so. We get peevish and irritable, cross, dirty and careless. Eat like beasts, our faces and hands begrimed with dirt and pine smoke and but little inclination to wash them or strength if we had. The weather has been so cold and we with so little wood that we cannot wash but on warmer days. Yesterday being quite a pleasant one all hands shaved up and took a good wash. This morning it stormed some. Am so hungry and faint that I shall do as upon many similar occasions, let it go. Well I'll dry up the writing till the day is passed, but I presume this will be made for the day.

Tuesday, 19 ⁻ Nothing to eat today but one pound of cornbread and one pint of U.S. coffee. I am too weak to make an effort to raise any extra and what I shall do I don't know. Well I guess Secesh is bound to starve us at all events.

Wednesday, 20 ⁻ One pound corn bread and half pound of Irish potatoes. I go to bed seemingly nearly famished with hunger. My appetite is craving and I can't get enough to gratify it. O my God what shall I do. And yet, I don't think, or can I make myself believe, but what I shall yet come out all right.

Jan. 21, 1864 ⁻ Was at first provoked by drawing first two ounces of rice and 1½ pounds corn bread, but agreeably disappointed when the best bean soup was added. I ate about 1/8 of my soup, traded the balance for an issue of rice and made my cup full of a splendid cup of rice. Let us see one pint beans, 1½ rice and 1½ pound cornbread is a very fair breakfast. I could live on such rations. But this rice is an extra furnished by our own government and when we shall get any more God only knows. Why don't we get more to eat?

Jan. 22, 1864⁻ Our fair continually decreasing and I am so weak I almost despair lest I may yield to the cravings of hunger and sink.

Friday, 23⁻ Nothing new from yesterday.

Saturday, 24⁻ Same only the weather is getting warm and springlike.

25th⁻ Sold my Masonic pin today for $10, Confed. $1.00 U.S. bought rice and molasses and went to selling them, made $8.00 (Confed.) first day and felt much encouraged. What was better our rations were nearly doubled and I went to bed with a full stomach and slept soundly, having left old Howard as he is called by all the boys bunking with our ―――― boys.

My Old Camp Complaint

26th - Hi ho got a letter from Eva, the first I have received since my capture. O bless you Eva, ever faithful and true. I'll try to be to you a good kind husband with God's permission. Weather very warm and springlike.

27th - Rice business dull today and did not make anything but kept my capital good and had enough to eat. That is more than I can do every day.

28th - Wound up the business today by selling what stock I had on hand and will try rations with a small addition. Forgot to mention the receipt of a letter from Capt. Dean[30] promising a box of provisions and money. The money is now due.

Friday, 29th - Wrote or rather sent letter to Capt. Dean and Eva, writing yesterday and day before. Gave note to Sergt. Hill concerning express boxes. Weather continues warm and springlike, very much so. Took a thorough wash today all over.

Saturday, Jan. 30 - Events transpired in the order recorded though I got the days of the week somewhat

---

[30] Capt. Charles W. Dean of the 6th Michigan Cavalry's Company I. He was not present at Charles Town, and had since been promoted to major of the regiment.

mixed. This morning the sky was overcast with clouds and the air raw and disagreeable. My old camp complaint[31] *is again around me. The news nearly nothing. Have been thinking seriously of writing to father for a box of provisions. Guess I will as soon as I can get some paper.*

Sunday, Jan. 31 ⁻ Nothing new. Our usual corn dodger, as we have long since denominated our ration of corn bread, with soup generally of rice or the "pea bean." They continue to hand out larger rations than formerly.

Monday, Feb. 1 ⁻ Will say tis the same as yesterday only the weather changed for the worse.

Tuesday, 2nd ⁻ And the same again today.

Wednesday, 3rd ⁻ And still another day the same, why encumber the pages with more only there is fairer weather again.

The pea bean soup Osborn often received wasn't exactly a dish like mother used to make. As fellow prisoner John Ransom described it, the soup "is made from the following recipe (don't know from what cook book, some new edition): Beans are very wormy and musty. Hard work finding a bean without from one to three bugs in it. They are put into a large caldron kettle of river water and boiled for a couple of hours. No seasoning, not even salt put into them. It is then taken out and brought inside. Six pails full for each squad ⁻

---

[31] Diarrhea

about a pint per man, and not over a pint of beans in each bucket. The water is hardly colored and I could see clear through to the bottom and count every bean in the pail. The men drink it because it is warm. There is not enough strength or substance in it to do any good. We sometimes have very good bean soup when they have meat to boil with it."

Other times the beans were cooked in the water used to boil the guards' bacon. A prisoner analyzed this version of the soup as 20 percent maggots, 30 percent beans and 50 percent water. The corn bread accompanying the soup was carried into the prison compound in vermin-infested blankets. Though they fared better when the rice, coffee and other delicacies were provided by the U.S. government, the prisoners often were so hungry they would struggle for potato peelings and bones thrown away by the guards.

Sgt. Daniel Gribban of the 132nd New York Volunteer Infantry, who survived his stay on Belle Isle, later escaped his captors and returned to Union lines. The New York Tribune reported Gribban's description of Belle Isle, and this account was reprinted elsewhere in the North: "The unfortunate soldiers looked more like skeletons than human beings. Hundreds were unable to walk, and those lay huddled together in groups until welcome death came and snatched them from their horrible sufferings. A case was known where seven dead men lay thus openly exposed for eight days without burial. What an awful sight this must have been for their famished comrades who tottered around, feeling that their end was fast approaching, without any earthly hope of reprieve or comfort.

"Another case was known when a hungry soldier,

being provided with beans, having had no nourishment for three days, swallowed the food ravenously, but his stomach being too weak was unable to hold anything, and the beans were cast up upon the ground. Another poor fellow, who was also dying from starvation, eat those very same beans and enjoyed them as a delicious morsel.

"A dog came into the lines one day and was instantly devoured. Shooting at the prisoners by the rebel guard was also a frequent occurrence. In one instance two soldiers were instantly killed and one wounded. These may be shocking facts to read, but it is well to know what our gallant soldiers have to suffer while in the hands of their remorseless foe."

A Confederate view of the compound's degenerating condition was offered by G. William Semple, surgeon in charge of the Belle Isle prisoners: "Into the camp containing an area sufficient for the accommodation of about 3,000 men have been crowded for many months past from 6,000 to 10,000 prisoners. To prevent escapes they have not been allowed to visit the sinks at night. These deposits of excrement have been made in the streets and small vacant spaces between the tents.

"The streets are so greatly crowded during the day as greatly to interfere with the working police parties, so that nearly the whole day is consumed by them in the imperfect removal of the filthy accumulations of the night. The whole surface of the camp has thus been saturated with putrid animal matter. Surrounded by such circumstances the prisoners have been totally careless of personal cleanliness.

"The rations now consist entirely of bread, rice, and

peas or beans. The bread is made of corn-meal, unsifted or bolted. Not separating the bran from the meal tends greatly to cause and continue the two diseases (diarrhea and dysentery) most prevalent among the prisoners. Many of them are destitute of blankets, having sold the articles lately furnished them by their Government. Under these circumstances, though they have been furnished with fuel, there has been great suffering from cold during the unusually cold weather of January and February, to which the brutal conduct of the prisoners in expelling their comrades from their tents at night has greatly added.

"To the crowded and necessarily filthy condition of the camp, the absence of personal cleanliness of the prisoners, the meager rations, and the effects of cold may be added the depressing effect of long continued confinement without employment, mental or physical, and with little hope of an early termination of the imprisonment, which together make up a sufficient sum of causes to amount to one-fourth the average number of prisoners in the camp. The diseases have been such, consisting principally of typhoid fever, diarrhea, dysentery, and catarrh and the diseases of the respiratory organs, as might be expected to result from the causes stated. But great as is this amount of disease, it is not greater than the average sickness among the U.S. troops in the field on the Atlantic coast, as reported by one of their own surgeons.

"Much difficulty has been experienced in procuring the regular attendance of the sick at sick-call. Patients who have been brought out on litters, unable to walk and greatly emaciated, who have never before attended sick-call, and several deaths have occurred in camp without the prisoners having been seen by or reported to a medical officer...

"The commanding officer and his subordinates have done all within their power to carry out [what has been] suggested toward the sanitary condition of the camp, but the number of officers and the guard and police seem too small to re-establish and enforce such system of police as would enable the medical officers to discharge their duties with as much efficiency and satisfaction as amongst our own troops in the field."

View of Belle Isle from the south side of the James River. The bridge to the island is in the foreground, and Richmond is in the background. (Massachusetts Commandery Military Order of the Loyal Legion and the U.S. Army Military History Institute.)

Nearer to an End

Thursday, 4th - Hereafter I shall not try to keep a daily record of events as this book is nearly full and I don't know how to keep another. Suffice it to say here that general prospects of our release do not increase except as time passing brings us nearer to an end - perhaps our own in time. I shall therefore wait until something new turns up and note it down as it occurs upon the line.

Osborn's penmanship, so neat and within the lines on the diary's pages at the beginning, had steadily deteriorated. As he filled more and more of the pages he began crowding the words together. At this point he was making use of every bit of space available, fitting in two lines of writing to each rule on the page.

On Feb. 6, he took a break from his diary entries, and instead penned a letter to his father back in Indiana.

Father, after so long a silence I write you again. I should have done so long ago but I had heard nothing from a single person in the North. Last week I received a letter from Eva saying you had written her & that "you thought a letter would not reach me." Tis a mistake. Write by all means, but briefly & nothing contraband. I am not well. Have not been since I came here never expect to be again, though I shall weather it through if I remain another year. We have been hoping all along to be released upon some terms. The day is now apparently as distant as ever. The soldiers, great numbers of them have been receiving Expressed boxes of necessaries. I have concluded the circumstances will justify me in asking you to put me up a box of substantial eatables, send to Edgerton or Antwerp & have it Expressed to me. I believe there is no expense to the sender or receiver. But what if there is, I need them & know you will send them. Let me suggest, Smoked ham 20 lbs. Maple sugar 15 lbs. cheese 10 lbs coffee 5 lb Dried Beef 25 lb Butter 10 to 15 lbs or in lieu of Butter Lard or Fried fat put up in water tight vessels, Tea 2 lbs a litttle pepper, mustard or any thing say Raddish roots. I (undecipheable) a large box, of course you can make it of any convenient quantities, only the box must not weigh to exceed 100 lbs.

You will watch the news, should an exchange be affected before you are ready to send the box, hold on to it. If not please send it to Sergt J.O. Coburn U.S. Prisoner of War, Richmond, Va (Belle Island prison) below. I will settle the bill at some future day. Write me a long letter to send in the box giving all news, & a brief one to send by mail addressed in the same manner. All boxes are inspected here in the presence of the owner. Any contraband matter renders the whole liable to confiscation. All Northern money is contraband except gold & silver. No money of any kind. Hoping to hear *from you soon but that it may be inside our own lines. I am as Ever,*

   Osborn

Written while nearing the brink of starvation, the letter reads like a desperate wish list of the most satisfying foods he can imagine, a hoping almost beyond hope for life-sustaining sustenance.

   Tuesday 9th (Feb.) ⁻ After several days of fair but cool weather and this one being spent as usual, all hands were greeted with surprise by a brass band striking the "Mocking Birds" at the front. All hands prophesied some good news in store for us, but it proved that they had only come over to have a drink with the lieutenant commanding here and they kept it up till the wee small hours of the morning.

   Wednesday, Feb. 10 ⁻ I am quite poorly today but finished another counterfeit Confed. bill. For a week passed this is the way I keep up extra rations. Pretty business for me making counterfeit money.

   Thursday, 11th ⁻ I felt so much worse today that I reported to the sick call. Stayed until I was tired and cold. No doctor came and I returned.

   Friday (Feb.) ⁻ Still failing. I went to doctor's call again and was unexpectedly ordered to gen. hospital, on the whole I was not sorry especially after being nearly all day in getting through. I found myself so completely exhausted that I could scarcely stand. Finding myself greeted as one of the sickest in consequence I took my bunk, kept it and suffered myself to be treated accordingly.

Prisoners tended to view the hospitals as a last resort. One historian wrote that Belle Isle's hospitals "were unfit for stables, and patients seldom admitted till they were ready to die." A March 8, 1864 report by Richmond medical director W.A. Carrington gave weight to the prisoners' concern. Carrington noted that of the 2,200 POWs in Richmond hospitals in February, 590 died.

Osborn was admitted to Confederate General Hospital No. 21, at the northwest corner of Richmond's 25th and Cary streets.

The 132nd New York's Sgt. Gribban was a wardmaster at this hospital, and the New York Tribune described his account of the facility: "The medical treatment given our poor fellows at the hospital, by Drs. Semple and Gibbs, is described as of the most barbarous character. Amputations were performed indiscriminately, whether it became necessary or not. Men were allowed to die of hemorrhage when a little exertion on the part of the physicians would have saved their lives. Several were known to bleed to death at night, because Dr. Gibbs did not feel like getting out of bed. This man had a perfect mania for amputating limbs. He would scarcely allow a soldier to go from the hospital without he left behind him part of his body. A number died under the brutal, and at all times clumsy operations."

Nelson D. Ferguson, surgeon of the 8th New York

Cavalry, spent four days at Hospital No. 21 assisting with the care of sick and wounded prisoners. Although he described the hospital buildings in Richmond as tolerably clean and "well suited for such purposes, being large, convenient and well ventilated," the bedding in Hospital No. 21 he found to be "very dirty. The covering was entirely old dirty quilts. The beds were offensive from the discharges from wounds and secretions of the body, and were utterly unfit to place a sick or wounded man on."

"I consider the nourishment and stimulation they received entirely insufficient to give them a proper chance for recovery," Ferguson said of the hospital's patients. "I am surprised that more do not die."

De Witt Peters, an assistant surgeon who treated released Belle Isle prisoners at Jarvis Hospital in Baltimore, reported that patients at Hospital No. 21 had an advantage over the prisoners on Belle Isle: "They were allowed to buy a loaf of bread the size of a man's fist, for which they paid five or six dollars in Confederate money."

As Jonathan Wilkins, surgeon in charge of Hospital No. 21, reported to Medical Director W.A. Carrington on March 7, "The daily list of deaths is regulated by the number admitted each day from Belle Isle. During the past month twenty-five cases died before they had been in the hospital twenty-four hours. It is so common an occurrence for the patients sent from Belle Isle to be speechless or delirious and unable to give their names, &c., that I have requested the surgeon in charge, in addition to the list forwarded by the conductor of the ambulances, to pin their names, companies, and regiments of desperate cases on the lapel of their coats.

"The majority of cases die of chronic diarrhea. During the past month 337 cases suffering with this disease were admitted. The deaths from this disease during the month sum up to 265. Of typhoid fever cases for the last month 64.5 percent have died; from diarrhea 59.7 percent.

"The commissary department for five weeks has not been able to furnish me with flour. The meal furnished in lieu thereof is ground with the husk and will produce diarrhea. I have ordered it to be sifted, but it is ground too fine to separate the husk from the meal.

"The medical purveyor does not furnish the hospital with a sufficient quantity of medicines. I made a requisition on the 1st of March, which has not been filled as yet. I would be most happy to receive suggestions from you in the treatment of diarrhea. I believe the medical officers have tried all known and approved remedies for the disease.

"In the case of other diseases, as pneumonia, &c., they generally occur in constitutions already enfeebled by diarrhea, and are generally in the second stage when admitted. In conclusion, the prominent character of all cases is emaciation."

Chronic diarrhea was Osborn Coburn's diagnosis.

I will remark here that I was much surprised from the start in the management of this prison hospital. Our own boys are far more attentive and kind than I had been led to anticipate by reports, while it is the infernal Reb authorities that do nearly all the stealing and swindling. It seems as if it could be, that at the proper time God will curse this sin polluted people. They are now over run with lice. Will not all the plagues of Egypt be visited upon them? My own nurse is a true Buckeye boy (1st. O. Cav.) and has already shown me some attention, God help him. In extras or luxeries, I have received a few butter crackers, tamarinds, Java coffee and beef tea from the extract, all from our own good government. The Rebellion shows us no such favors at her own expense.

Saturday, 13th - About the same as yesterday.

Sunday, 14th, Valentine's day - And how I would like to be exerting my ingenuity upon the usual love epistles, but the privilege is denied me and I must be content. I had all the coffee I had ought to drink today though I could have relished more, about three pints in all.

Tuesday, 16th - I notice for some reason our luxuries are all cut off so we are down to corn dodger and soup or a small morsel of meat. Last night I got about two gills of white beans, this morning nothing but a small piece of corn bread. Think of that for a sick man in what is pretended a civilized country.

Memoranda. Complaints have been made as to our treatment both here and on the island, about the first of the month some persons with the med. director went to the island and had the rations increased. The Reb. congress then orders an investigation which occurred I think about the sixth, make a report to congress who report on the 18th that the charges are without foundation. On the tenth our rations were as small as ever and thus stands their record before the world.

Sunday, 21st (Feb.) – *Today I wrote Hawley[32] and Eva. I sent to Hawley for a box of eatables. Wonder if it will ever come. Have been gaining in all respects but strength. Faith and courage good. Have been bunking for several days with a young man by the name of Nichols. Rather an intelligent young man and will remember him. Our rations are not permanently reduced. My nurse favors me a little and I am doing quite well.*

Monday, Feb. 22 – *Nichols was transferred to another hospital this morning. He was very kind.*

Here four pages of the diary are gone.

Wednesday, March 2, 1864 – *The morning opens bright and clear but cool after a two days rain storm. I feel bright as a dollar. Got up, made up report, toasted my bread, ate my breakfast, ate heartier than for several days, felt far less inconvenienced from it. Am not very tired as I have generally felt.*

Memoranda. If they destroy that property by G⁻d I'll kill every d⁻d in Richmond. This about the first of February 1864.

That odd message was Osborn's last diary entry.

---

[32] Hawley Thomas, Mattie's brother and Osborn's brother-in-law.

Postwar CDV photograph of Cynthia D. (Harvey) Thomas, wife of J.O. Coburn's brother-in-law Hawley Thomas. The couple were married Nov. 14, 1861, and in 1864 were residing in Jackson County, Mich.

A postwar CDV photograph of J.O. Coburn's niece and nephew, children of Cynthia and Hawley Thomas. Lillian May Thomas, right, was born Feb. 6, 1863, and Grantie Heman Thomas, left, was born March 14, 1864.

Despite his high hopes Osborn didn't come out all right. On March 8, 1864, he died in Hospital No. 21. Diarrhea was attributed as the cause. He was now just a statistic, one among 40,000 Union men who lost their lives as Confederate prisoners.

Big Rapids received its first intimation of Osborn's fate through a letter from a former Belle Isle captive, an unidentified member of the 3rd Michigan Cavalry. A portion of the letter was printed in the July 1 *Mecosta County Pioneer*: "Coburn, I think, must be dead; for the last time I saw him he was very sick and weak, reduced to a skeleton. He was sent to the Hospital in Richmond from Belle Isle, and if he was alive, he would have been sent through to our lines, as the rebs were sending none but the sick at that time..."

Even in death Osborn was denied a return home. He was buried in Richmond, and now lies among fellow prisoners in the Richmond National Cemetery.

But the diary did make it north, thanks to fellow prisoner Albert S. Patrick of New York, a hospital steward who was at Osborn's deathbed. Patrick saw to it that the diary found its way to Osborn's father.

A funeral service for Osborn was conducted May 1, 1864, near his father's home at Coburn's Corners in DeKalb County, Ind. An account of the service, including part of a letter Patrick wrote to the elder Coburn, was contained in the Aug. 27, 1864 issue of The New Era, an Auburn, Ind., newspaper. "Your son fell a victim to the inhuman cruelty and barbarous treatment of our captors," Patrick wrote, "and I do not overstate the truth when I say that he literally starved to death on Belle Isle. He came to the prisoners' hospital on the 14th of February, sick with chronic diarrhea. I was acting Stewart at the time, and when he came in, noticed he was a man of more than ordinary intelligence, and soon made his acquaintance. I furnished him every comfort possible under the circumstances, and he so far recovered as to come down stairs and try to do my writing. I felt greatly encouraged in regard to his case, and cheered him all I could, but three days had not passed before he relapsed, and in forty-eight hours was dead. His system was completely worn out and nature could not rally. To give you some idea of how fearfully emaciated he was, I will tell you that his ankle was far larger than his thigh, and his hip bones protruded through the skin...His mind was clear up to his latest breath and he conversed in regard to his approaching end, calmly, and I never before saw a man to whom death had so few horrors; and whenever I think of him, I pray my last end may be like his."

Graves of Union soldiers on Belle Isle. (Massachusetts Commandery Military Order of the Loyal Legion and U.S. Army Military History Institute.)

Epilogue

Two officers and 25 enlisted men from Company I of the 6th Michigan Cavalry were captured at Charles Town. Of the enlisted men, 17 died in captivity, no further record is found for one, and only seven survived. Both the officers survived, being held in different accommodations than the enlisted men.

Cpl. Edwin Beckwith died at Andersonville, Ga., on May 31, 1864. He is buried in the Andersonville National Cemetery. At age 23 Beckwith enlisted in the 6th Michigan Cavalry Sept. 2, 1862, at Holly, Mich., and he began service as a corporal.

Pvt. Anderson Bunch died Aug. 29, 1864, in prison at Augusta, Ga.. He is buried in the Raleigh, N.C., National Cemetery. At age 28 Bunch enlisted in the regiment Aug. 29, 1862, at Stronach, Mich.

Theodore Burdick, a bugler, died at Andersonville on Aug. 27, 1864. He is buried in the Andersonville National Cemetery. At age 20 Burdick enlisted in the 6th Michigan Cavalry Aug. 30, 1862, at Stronach, Mich. He was wounded in action at Charles Town before being taken prisoner.

Quartermaster Sgt. Jacob Osborn Coburn died March 8, 1864, at Hospital No. 21 in Richmond, Va., of chronic diarrhea. He was buried in Richmond, likely in the city's poorhouse cemetery, and in 1867 was among the Union soldiers reinterred in the Richmond National Cemetery, where he lies in grave 584. At age 30 Coburn enlisted in the 6th Michigan Cavalry on Sept. 8, 1862, at Leonard, Mich., and began service as a corporal.

Pvt. Luman A. Cutler died Nov. 26, 1863, as a POW at Richmond, Va. He is buried in Richmond National Cemetery. At age 20 Cutler enlisted in the regiment at Otto, Mich., on Aug. 30, 1862.

Pvt. John Deits died at Andersonville on March 14, 1864. He is buried in Andersonville National Cemetery. At age 21 Deits enlisted in the 6th Michigan Cavalry on Sept. 3, 1862, at Manistee, Mich..

Pvt. Charles Franklin died Jan. 14, 1864, as a POW at a Confederate military hospital in Richmond, Va. He is buried in the Richmond National Cemetery. At age 33 Franklin enlisted in the regiment Sept. 5, 1862, at Leonard, Mich.

Cpl. George C. Hoag died Dec. 24, 1863, as a POW in Richmond, Va. He is buried at Richmond. At age 25 Hoag enlisted in the 6th Michigan Cavalry Sept. 3, 1862, at Manistee, Mich.

Pvt. Edwin Nicholson died April 12, 1864, at Andersonville. He is buried at Andersonville National Cemetery. At age 23 Nicholson enlisted in the regiment Sept. 6, 1862, at Leonard, Mich.

Pvt. Levi Orner died Feb. 28, 1864, as a POW in Richmond, Va. His place of burial is not listed. At age 28 Orner enlisted in the 6th Michigan Cavalry Sept. 16, 1862 at Holly, Mich.

Pvt. John Paradise died March 2, 1864. He is buried in the Marietta, Ga., National Cemetery. At age 37 Paradise enlisted in the regiment Sept. 3, 1862, at Manistee, Mich.

Pvt. Thomas Parish died June 1, 1864, at Andersonville. He is buried at the Andersonville National Cemetery. At age 18 Parish enlisted in the 6th Michigan Sept. 16, 1862, at White Lake, Mich.

Pvt. Robert H. Payne died June 14, 1864, at Andersonville. He is buried in the Andersonville National Cemetery. At age 31 Payne enlisted in the regiment Aug. 25, 1862, at Tyrone, Mich.

Pvt. John Pfieffer died June 26, 1864, at Andersonville. He is buried in Andersonville National Cemetery. At age 25 Pfieffer enlisted in the 6th Michigan Cavalry Aug. 29, 1864, at Stronach, Mich.

Pvt. Charles Riley died July 20, 1864, at Andersonville. He is buried in Andersonville National Cemetery. At age 19 Riley enlisted in the regiment Sept. 5, 1862, at Manistee, Mich.

Pvt. Eleazar H. Thatcher died July 18, 1864, at Andersonville. He is buried in Andersonville National Cemetery. At age 18 Thatcher enlisted in the 6th Michigan Cavalry Sept. 8, 1862, at Fenton, Mich.

Pvt. Joel B. Way died in August 1864 of disease as a POW at Augusta, Ga.. His place of burial is not listed. At age 25 Way enlisted in the regiment Sept. 6, 1862, at Fenton, Mich.

For one 6th Michigan Cavalry Company I soldier captured at Charles Town, **Pvt. Frank Weinrich**, no further record is found. At age 26 he enlisted in the regiment Sept. 1, 1862, at Stronach, Mich.

Of the seven 6th Michigan Cavalry Company I enlisted men who survived their imprisonment, most endured the prisons at Richmond, Andersonville and later other locations before their release or exchange.

Cpl. James Connolly returned to the regiment June 10, 1865. He was mustered out of the service Nov. 24, 1865, at Fort Leavenworth, Kan. At age 24 Connolly enlisted in the regiment Nov. 14, 1862.

Pvt. John M. Crawford returned to the regiment to be promoted to corporal June 1, 1864, and to sergeant Feb. 1, 1865. He was mustered out of the service Nov. 24, 1865, at Fort Leavenworth, Kan. At age 21 Crawford enlisted in the regiment Sept. 10, 1862, at Manistee, Mich.

Pvt. John Fenning returned to the regiment June 10, 1865, and was mustered out of the service Nov. 24, 1865, at Fort Leavenworth, Kan. At age 27 Fenning enlisted in the 6th Michigan Cavalry Sept. 20, 1862, at Leonard, Mich.

Pvt. Moses Laurent was discharged from the service June 29, 1865, at Annapolis, Md. At age 22 Laurent entered the regiment Aug. 29, 1862, at Stronach, Mich.

Pvt. William Shaw returned to the regiment June 10, 1865, and was mustered out of the service Nov. 24, 1865, at Fort Leavenworth, Kan. He survived two stints as a POW, as he was first captured June 11, 1863, at Seneca, Md., and returned to the regiment Sept. 1, 1863. At age 18 Shaw enlisted in the regiment Aug. 29, 1862, at Stronach, Mich.

Pvt. William Teggerdine was discharged from the

service June 2, 1865, at David's Island, N.Y. At age 19 Teggerdine enlisted in the regiment Sept. 16, 1862, at White Lake, Mich.

Pvt. Orrin E. White returned to the regiment June 10, 1865, and was mustered out of the service Nov. 24, 1865, at Fort Leavenworth, Kan. He was appointed corporal dating from Feb. 1, 1865. At age 18 White enlisted in the regiment Sept. 8, 1862, at Fenton, Mich.

1st Lt. Robert A. Moon returned to the regiment May 9, 1865, and was mustered out of the service Nov. 24, 1865, at Fort Leavenworth, Kan. At age 30 Moon enlisted in the regiment Sept. 1, 1862, at Grand Rapids, Mich., and he entered the service as a first lieutenant.

2nd Lt. Malcom M. Moore returned to the regiment April 1, 1865, and was mustered out of the service Nov. 24, 1865, at Fort Leavenworth, Kan. At age 19 Moore enlisted in the regiment Aug. 28, 1862, at Grand Rapids, Mich., and he entered the service as supernumerary second lieutenant. He was commissioned second lieutenant March 16, 1863 and first lieutenant May 23, 1865. Moore also was promoted to brevet major of U.S. volunteers March 13, 1865, for gallant and meritorious service during the war.

The 6th Michigan Cavalry Company I commander, **Captain Charles W. Dean**, was not present at Charles Town. At age 25 Dean enlisted in the regiment Aug. 25, 1862, and he was commissioned captain Oct. 13, 1862. He was promoted to major Nov. 11, 1863, was wounded at Shepherdstown, Va., on Aug. 25, 1864, and resigned from the service Jan. 5, 1865.

Members of other units mentioned in Osborn Coburn's diary include:

Brig. Gen. Neal Dow, who was commissioned colonel of the 13th Maine Volunteer Infantry on Nov. 23, 1861, at age 57. He commanded the Department of the Gulf's District of West Florida from Oct. 2, 1862, to Jan. 23, 1863, and the First Brigade, Second Division of the XIX Corps from Feb. 26 to May 27, 1863, when he was twice wounded at Port Hudson, La. While recovering he was captured and sent to Libby Prison in Richmond, Va. On March 14, 1864, he was exchanged for Confederate Brig. Gen. W.H.F. "Rooney" Lee. Gen. Dow resigned Nov. 30, 1864, due to poor health.

1st Sgt. Asa Howard. He enlisted in the 20th New York State Militia, known in the Federal service as the 80th New York Volunteer Infantry, on Oct. 5, 1861, at Kingston, N.Y. Captured at Gettysburg on July 1, 1863, he died as a prisoner March 2, 1864, at Augusta, Ga. Howard was promoted to third sergeant June 6, 1862, and to first sergeant Oct. 1, 1862, and was reduced to the ranks on July 31, 1863, while a prisoner.

Sol Miller, who could not be positively identified.

William Perry Montonye enlisted Feb. 22, 1862, in Company E, 3rd Michigan Cavalry. Captured Aug. 19, 1863 while on a raid near Grenada, Miss. Montonye was held for 11 months in prisons at Columbus, Miss., Richmond, Va., Andersonville, Ga., and Savannah, Ga., before being exchanged. He was discharged from the service at Baton Rouge, La., in June 1865. He returned to Big Rapids, where he still resided in 1884.

Cpl. Daniel Will, Company C, 100th Ohio Volunteer Infantry. Captured Sept. 8, 1863, at Limestone Station,

Tenn., he was approximately 23 while at Belle Isle. He survived Belle Isle and Andersonville prisons, and returned to his company June 14, 1864. Daniel Will survived the war and later moved from Bryan, Ohio, to California, and he died near Los Angeles on Dec. 13, 1914, at age 76.

Sgt. Turner M. Wynn, Company C, 100th Ohio Volunteer Infantry. Captured Sept. 8, 1863, at Limestone Station, Tenn., he was approximately 21 while at Belle Isle. He survived Belle Isle and Andersonville prisons, but died Nov. 27, 1864, in the Confederate prison at Savannah, Ga.

Osborn's fiancé, Eva Aldrich, was living with her parents in Onstott, New York, at the time of his death. In 1865 Eva Aldrich, then 29, was living with her mother in Onstott, and had been appointed postmistress of the village replacing her father, who died in 1864.

On July 20, 1865 Eva was married to Selden R. Goddard, 26, wagon maker of Newfane, New York. Like Osborn Coburn, Goddard had served in the Union army, enlisting in August 1862 with Company K of the 152st New York Volunteer Infantry. He was a corporal, and was discharged from the New York regiment on Feb. 24, 1864, to become an officer with the U.S. Colored Troops. He resigned June 3, 1865, as a first lieutenant in the 30th U.S. Colored Troops.

In 1873 their four-year-old daughter died. Eva died Oct. 9, 1885, at age 50, leaving her husband and two children.

View overlooking the empty Belle Isle prison. (Massachusetts Commandery Military Order of the Loyal Legion and the U.S. Military History Institute.)

In 1867 J. Osborn Coburn's body was reinterred in Richmond National Cemetery, in what is today an unknown grave. Although initially the soldiers' graves there were marked with numbers, citizens of Richmond pulled up the wooden markers and used them for firewood. Author Don Allison is shown with a memorial stone to Coburn that he arranged to be placed in Richmond National Cemetery in 1997. (Photo by Diane Allison.)

In 1997 a ceremony was conducted to dedicate J. Osborn Coburn's memorial stone in Richmond National Cemetery. Above, author Don Allison, far right, delivers remarks. Below, Diane Allison prepares to place a wreath at the stone.

A Coburn family marker in the Coburn's Corners cemetery contains these memorial inscriptions for J.O. and Martha Coburn. (Don Allison photo.)

Appendix

Review by Judge-Advocate-General Joseph Holt, U.S. Army, of the proceedings of a court of inquiry, convened by Brig. Gen. Benjamin F. Kelley, commanding the Department of West Virginia, at the request of Col. Benjamin L. Simpson, Ninth Maryland Infantry, to investigate the circumstances attending the surprise and capture of Charlestown, W. Va., October 18, 1863.

* * * * * *

OPINION OF THE COURT.

The undersigned have obtained all the evidence in this case which has been attainable. They have carefully considered and compared it, and in obedience to the order convening them respectfully express their opinion to be:

That the surprise and capture of the greater part of the forces under the command of Col. Benjamin L. Simpson, Ninth Maryland Infantry, at Charlestown, Va., on the morning of the 18th day of October ultimo, were inevitable, because of the peculiar location of the place, which, surrounded by an open country for several miles on all sides, and approachable by a large number of roads from all directions, was easy to be flanked and surrounded, and because of the superior force by which it was attacked and of the inferior force for its defense.

The rebel attacking force was 2,000 men, with six pieces of artillery. The defending force, under the command of Colonel Simpson, consisted of 375 infantry and 75 or 80 cavalry. Although the pickets of Colonel Simpson were posted not so far out as at first view might seem to have been desirable, yet when the smallness of his force and the circuit of his picket lines, extending as they were established 3 miles, are taken into account, it is not perceived how he can be justly held blamable for not extending them farther. The proof shows that General Lockwood when in command directed one of the posts to be drawn in from what he considered its too great exposure. That they were driven in simultaneously and rapidly was due to the location of Charlestown, and the fact that the enemy had availed of it to post his forces around it on all sides during the darkness of the night of the 17th, and so to be ready for attack on each picket post at the same moment, which it may well be inferred had been previously fixed on.

Colonel Simpson's infantry force consisted of part of the Ninth Maryland Volunteers, which had been in the service but two months and had not been under fire before. The exposure to an artillery fire as described in the testimony, whilst the enemy approached their position by cross and by streets, keeping concealed from view and refusing fair combat, was well calculated to demoralize new troops.

Still a very painful feature of the affair was the disorganized condition of the men from the time they left the yard of the court-house to the time of their capture. In such a case the undersigned are of opinion that the field and line officers should have enforced the orders given to form column and line respectively, and should have maintained military order and decorum and consequent efficiency, by such use of their side-arms upon their own men as might have been necessary to accomplish these results.

In such case the worst enemies to the whole theory and spirit of the Army Regulations are those men who refuse from any cause compliance with the first paragraph of the first article of those regulations, and such enemies should be met and subdued by all the means at command of their officers. The undersigned are not aware that the course of action indicated has been recognized as a general rule governing officers under like circumstances, and they cannot therefore feel justified in censuring Colonel Simpson or his officers for having failed to pursue a course consonant with their opinions but not demanded by a generally recognized rule, whilst they think it unfortunate that it did not occur to them to resort to it.

They entertain no doubt, however, that whatever might have been the action of Colonel Simpson and his officers in this respect, the result would have been the same. The capture was inevitable, from the superior force of the enemy and its success in having selected its positions without hinderance, because of the facilities before alluded to. All the facts and circumstances considered together, the undersigned think that no blame can justly be attached to Colonel Simpson on account of the surprise and capture of Charlestown, Va., on the 18th October last.

There is a feature of this transaction which the undersigned think ought not to be passed without notice. It is a clearly established fact that the rebel forces, commanded by General Imboden, did, on the occasion referred to, shell Charlestown whilst it was occupied by women and children without allowing time for their possible removal from the localities shelled. And it is notorious that the population of Charlestown is composed, almost wholly, of active, or warmly sympathizing, co-laborers and friends of General Imboden in the work of rebellion. And this was done by a general with a force of 2,000 men and six pieces of artillery attacking a force of 450 men with not one piece of artillery.

It is irresistibly inferable that the object of the rebel commander was the capture of the force, so greatly inferior to his own, before that force could receive re-enforcements to put it on an equality with his, and so have opportunity for a fair, equal, and manly fight; and that, to the accomplishment of this object, he did not hesitate to sacrifice the safety of the women and children of the ⁻ by him and them ⁻ claimed Southern Confederacy.

In fact, within a short period after the accomplishment of this feat, the entire force of General Imboden retreated rapidly before the charge of Major Cole's battalion of about 300 men, which had come up to Colonel Simpson's relief; and before that battalion and one regiment of infantry (the Thirty-fourth Massachusetts) and one battery of artillery (Miner's), which shortly after came up, all under command of Col. George D. Wells, commanding First Brigade, and numbering less than 700 men all told, he continued industriously his retreat, with his 2,000 men and six pieces of artillery, for more than 9 miles, and until the pursuing force was recalled.

The loss of Colonel Simpson's regiment at Charlestown was 2 men killed, 2 officers and 7 men wounded, 16 officers and 340 men captured; 4 wagons, 2 ambulances, and 20 horses, together with some arms, and the ammunition on the persons of the men, about 60 rounds each.

There was but a small quantity of quartermaster's or other stores on hand.

Col. 1st Maryland P. H. B. Vols., President Court of Inquiry.
   F WM P. MAULSBY,

   RANK A. ROLFE,
   Major First Massachusetts Heavy Artillery.

   W. B. CURTIS,
   Major Twelfth Regiment West Virginia Volunteer Infantry.

   [Indorsement.]

   HEADQUARTERS DEPARTMENT OF WEST VIRGINIA,
   Cumberland, Md., December 22, 1863.

The within proceedings of a court of inquiry convened at Harpers Ferry, Va., to inquire into the facts and circumstances connected with the surprise and capture, on the 18th day of October last at Charlestown, Va., of the forces under the command of Col. Benjamin L. Simpson, Ninth Maryland Volunteers, are respectfully forwarded for the consideration of the honorable Secretary of War.

After a careful perusal of the testimony adduced, I cannot concur in the opinion expressed by the Court that no blame can justly be attached to Colonel Simpson on account of the surprise and capture of Charlestown, or that the capture was inevitable.

I am of the opinion that Colonel Simpson was derelict in suffering his command to be surprised, and that he could and should have maintained himself in his position until he could have been relieved by the forces at Harper's Ferry, which were started to his assistance as soon as the cannonading of the enemy was heard at that place. I therefore, respectfully recommend that Col. Benjamin L. Simpson, Ninth Maryland Volunteer Infantry, be dismissed from the service of the United States.

B. F. KELLEY,
Brigadier-General.

REVIEW.

JUDGE-ADVOCATE-GENERAL'S OFFICE,

January 23, 1864.

The within is a record of the proceedings of a court of inquiry convened at Harper's Ferry on the 13th of November at the request of Colonel Simpson, of the Ninth Maryland Volunteers, to investigate the facts and circumstances connected with the surprise and capture of Charlestown and the forces occupying it on the 18th of October, and to give an opinion in the case.

From the evidence adduced the following facts appear:

The village of Charlestown is built upon uneven ground, surrounded by wooded hills and valleys. There are eight roads leading to it from different directions, all connected by another, which at some distance from the town completely encircles it. On the 20th of August Colonel Simpson was ordered by his brigade commander to proceed with the forces under him from Loudoun Heights and encamp in the woods on the east side of the village.

After his arrival he received orders and instructions, some written, some verbal, all of which do not appear in evidence from the fact that some of the originals and the books in which they were copied have been lost or captured. The general tenor or substance of all these material to the case in question were presented for the consideration of the Court. Up to the date of the capture they appear to have been carried out promptly as far as practicable, and the result communicated by Colonel Simpson to his brigade and division commanders.

Colonel Simpson's command consisted of portions of seven companies of the Ninth Maryland Volunteers, amounting to about 356 men and a cavalry force of about 80 men.

His instructions from his brigade commander were that the holding of Charlestown itself was of no more importance than as though it was an open plain, but to be vigilant in scouting the country, watching the movements of the enemy in his front, to engage and whip any force that came within his reach that was not too strong for him, and if attacked by superior numbers to retire.

From about the middle of September there appears to have been a force of rebels in that vicinity, which for some time were believed to be small, under the command of Major White. About the 1st of October it was rumored that an advance of a part of Lee's army in that direction might be expected, and that unusual vigilance was necessary. On the 7th of October he was informed by Colonel Wells, commanding brigade, that Imboden was said to be preparing to make a raid on the railroad between Martinsburg and Harper's Ferry.

On the 13th of October he was directed to send a scouting party to Berryville, and one, consisting of 10 men, to follow the summit of Blue Ridge toward Front Royal. On the same day he informed Colonel Wells that he sent a force to Berryville, which drove in the enemy's pickets (supposed to be of White's battalion), but being too small to make an attack, returned; and that they could hear of no force approaching up the valley; and that the scout ordered to Blue Ridge had returned, being unable to proceed on that road; and he asked if they should endeavor to go forward by any other route.

On the 14th it also appears that he had his wagons packed, and all preparations made to fall back to Harper's Ferry, but received an order from General Sullivan to wait till he was attacked. During the day he sent word to Colonel Wells that he had scouted every road leading from Charlestown, had found a force at Berryville, supposed to be White's, and a small force at Smithfield, but had gained no information of any heavy force in the valley. That he was informed that it was the intention to annoy his pickets that night; therefore he had increased their numbers, and would advance his posts on some of the roads, and have a company of men under arms to re-enforce any point that might be attacked.

On the 15th he was informed by Colonel Wells that it was reported in Harper's Ferry that Imboden was at Berryville with "something of a force," and that a part of it was being sent around between Charlestown and the river, and that the matter should be looked into. Colonel Wells closed his communication by saying that the alarm about an approach up the valley seemed to have been without foundation, and that Colonel Simpson could resume the even tenor of his way at Charlestown the same as before. On the same day Colonel Wells further informed him that he had sent a detachment of cavalry on the road from Martinsburg to Winchester and Berryville, with orders to convey any information they might obtain to Charlestown.

On the 17th Colonel Simpson informed Colonel Wells that a detachment of cavalry came into Charlestown the night before, being unable to reach Berryville on account of White's occupation of it. That an officer of Cole's battalion had arrived, having been wounded in a skirmish, and that he represented that no other force was in the valley but Imboden's; and further that he had captured one of White's men, who informed him that White's battalion, consisting of 150 cavalry and 80 dismounted as infantry, had been at Berryville, had reconnoitered the position at Charlestown, and had found it too strong for them; and that White had moved away, and Imboden was 8 miles above Winchester with 800 men.

At half past 5 o'clock on the morning of the 18th, it appears that the enemy simultaneously attacked and drove in the pickets on the different roads about Charlestown, and planted one battery north and one south of the town, and sent (under a flag of truce) a demand for a surrender. This being promptly refused, another flag of truce was immediately sent with directions to remove the women and children from the vicinity of the court-house and jail. A few minutes after this message had been given to the officer who was ordered out to meet the bearer of it, the shelling commenced.

It is shown that when the alarm was given by the attack on the picket Colonel Simpson posted his men in the court-house, the jail, and another building, making loop-holes for muskets. The enemy's batteries were so planted behind buildings at a distance of two or three hundred yards that their fire was very effective, while the cannoneers were completely protected from the musketry of Colonel Simpson's force. The first shot fired struck the court-house, and several others followed killing and wounding several men and an officer.

Colonel Simpson then ordered the men to evacuate the buildings and form column by company in the street, and ordered all the cavalry force, about 80 men, under Lieutenant Moon, to reconnoiter the roads and find the weak points of the enemy. This officer led his men out on the road leading to Harper's Ferry, and finding the enemy ordered a charge, which would seem to have been improper, and resulted in the killing or capture of all but the lieutenant and 17 men. These returned and reported the fact.

Colonel Simpson had meantime learned the force of the enemy and the number of pieces of artillery, and his men, who had been organized but two months and had never been in action before, became panic stricken and could not be kept in line, broke in confusion, destroyed their arms, and were very soon all in the possession of the enemy. Colonel Simpson is stated to have been remarkably cool, and at the head of the column, assisted by his officers, by commands and threats endeavored to rally his men that an orderly retreat might be effected. After several attempts being unsuccessful, the men having scattered in all directions, he with what officers were mounted struck off through the fields and escaped.

The opinion expressed by the Court was, that under the circumstances Colonel Simpson should not be held blamable. They state the grounds of their opinion very elaborately, as will be seen on pages 49 to 54 of the record, to which attention is invited. General Kelley forwards the record, stating that he does not concur in the opinion of the Court, but believes that Colonel Simpson was derelict in allowing himself to be surprised, and that he should have maintained himself in his position until the re-enforcements reached him which started from Harper's Ferry as soon as the cannonading of the enemy was heard. He therefore recommends that Colonel Simpson be dismissed the service.

This recommendation is not concurred in. After a disaster of this kind has occurred, it is much less difficult for a military commander to review the details, and remark what should or what should not have been done, than for a subordinate to have anticipated the strength, position, and design of the enemy, and to successfully have met or withstood their attack.

General Kelley simply expressed the opinion that Colonel Simpson was derelict, and should have maintained his position until relieved. If he had made any suggestions or stated any facts which were not presented to the Court for their consideration, or which seem to have been overlooked, the proper course would be to direct a trial by court-martial, rather than to order a summary dismissal in the face of a favorable opinion expressed by a board of competent officers. It is believed that all the facts in this case were fully inquired into and a just decision arrived at, and no further action seems to be called for.

J. HOLT,
Judge-Advocate-General.

[Indorsement.]
WAR DEPARTMENT,
[March 22, 1864.]

Respectfully referred to the Adjutant-General.

Finding of Court of Inquiry in the case of Colonel Simpson is approved. This officer will be released and ordered to duty if under arrest.

By order of the Secretary of War:

ED. R. S. CANBY,
Brigadier-General and Assistant Adjutant-General.

(Colonel Simpson was mustered out of the service with his regiment on Feb. 23, 1864.)

Notes and Bibliography

Editing of Osborn Coburn's diary was kept to a minimum. Spelling was not changed from the transcription, except in cases of obvious typographical errors. Punctuation and paragraph breaks, however, were added for clarity.

Following are notations for souces used in individual chapters.

Steady Watch of the Sentinel,
Chapter 1

Accounts of the Oct. 18, 1863 battle at Charles Town, W.Va., and the beginning of Osborn Coburn's trek south were taken from "The War of the Rebellion: A Compilation of the Official Records of the Union and Confederate Armies" (hereafter "Official Records"), Government Printing Office, Washington, D.C., Series I, Vol. XXIX, Part I, pages 485-492 and 1010-1015; and "A Compendium of the War of the Rebellion" by Frederick Dyer, The Dyer Publishing Company, Des

Moines, Iowa, 1908, Part II, page 977, and Part III, pages 1236 and 1273.

Descriptions in the book's opening paragraphs not taken from Osborn Coburn's diary or the "Official Records" accounts just cited were drawn from elements common to prisoners' experiences as outlined in several publications, including: "Andersonville Diary, Escape, and List of the Dead", John L. Ransom author and publisher, Auburn, N.Y., 1881, chapter I; "The Smoked Yank", Revised Edition, by Melvin Grigsby, copyright 1888, chapters VII, VIII and IX ; "Andersonville, A Story of Rebel Military Prisons", by John McElroy, published by D.R. Locke, Toledo, Ohio, 1879, chapters IV, V and VI; "In and Out of Rebel Prisons" by Alonzo Cooper, R.J. Oliphant printer, Otswego, N.Y., 1888, chapter VI; "Life in Rebel Prisons" by Robert H. Kellogg, title page with publication date and publisher missing, chapter I.

Background information on Charles Town was obtained by a personal visit on Aug. 9, 1995; and "A tour Guide to the Civil War" by Alice Hamilton Cromie; Quadrangle Books, Chicago, Ill., 1965; page 376.

A Man of Many Talents,
Chapter II

An account of the 6th Michigan Cavalry Company I's skirmish at Seneca Mills, Md., was taken from "Official Records", Series I, Vol. XXVII, Part II, pages 786-787, and Part III, pages 50 and 95.

Biographical information on Osborn Coburn, his wife, Martha, his brother-in-law Hawley Thomas, his law partner C.C. Fuller and other mentioned in this chapter were taken from his service records in the National Archives in Washington, D.C.; "History of DeKalb County Indiana," Inter-State Publishing Company, Chicago 1885; Old Settler's Edition of the St. Joe, Indiana News, June 21, 1905; Defiance County, Ohio, marriage records; Magistrates Qualifications 1824-1879 Williams County, Ohio, compiled by Richard L. Cooley, Williams County Historical Society, Montpelier, Ohio, 1986; Williams County, Ohio, deed records; "Historical Collections, Collections and Researches Made by the Michigan Pioneer Historical Society", Vol. XXX, Wynkoop Hallenbeck Crawford Company, State Printers, Lansing, Mich. 1906, page 28; "History of Jackson County, Michigan", Vol. II, Inter-State Publishing Co., Chicago, Ill., 1881, pages 1111-1112; and "Portrait and Biographical Album, Mecosta County, Michigan", Chapman Brothers, Chicago, Ill. 1883, pages 573 and 590.

Coburn's life in Williams County, Ohio, is most specifically taken from a summary of diaries kept in the 1850s by Osborn and obtained by the late Genevieve Eicher of Napoleon. The summary of the diaries appeared in the Farmland News of Northwest Ohio in six weekly segments, beginning Dec. 3, 1986.

Much of the information pertaining to events in Mecosta County, Mich., came from the weekly *Mecosta County Pioneer* newspapers of April 17, 1862, through Sept. 25, 1862. Some biographical information on Eva Aldrich was outlined in a letter from the Niagara County, N.Y., Office of Historian, dated Nov. 20, 1998.

A description of the state of affairs in Michigan during the fall of 1862, when the Sixth Michigan was formed, is found in "Zachariah Chandler: An Outline

Sketch of His Life and Public Services" by the Detroit Post and Tribune, Detroit, Mich., 1880, chapter XIV.

On to Richmond
Chapter III

Osborn Coburn's letters quoted in this chapter appeared in the *Mecosta County Pioneer* editions of Oct. 16, 1862; Jan. 15, 1863; July 2, 1863; and Aug. 27, 1863.

Facts on formation and movements of the 6th Michigan Cavalry were taken from "Official Records," Series I, Vol. XXVII, Part I, page 489, and Part III, pages 524, 538 and 544-545, and Vol. LI, Part I, page 1069; "Record of Service of Michigan Volunteers in the Civil War 1861-1865", Vol. 36, published by authority of the Senate and House of Representatives of the Michigan Legislature, pages 1-10; "Michigan in the War", Revised Edition, compiled by Jno. Roberts, W.S. George & Co., State Printers and Binders, Lansing, Mich., 1882, pages 573-595; and "Cavalryman with Custer" by J.H. Kidd, Bantam Books, New York, 1991 edition. Coburn's letters and other information published in the *Mecosta County Pioneer* editions of Sept. 18, Oct. 16, Oct. 23 and Nov. 6 of 1862 and Jan. 15, May 28, July 2 and Aug. 27 of 1863, also were utilized.

A description of the state of affairs in Michigan during the fall of 1862, when the Sixth Michigan was formed, is found in "Zachariah Chandler: An Outline Sketch of His Life and Public Services" by the Detroit Post and Tribune, Detroit, Mich., 1880, chapter XIV.

Look and Listen, Wish and Wonder, Chapter IV

Coburn's letter to his father announcing that he is a POW is contained in "Some Pioneer Settlers of Coburntown," compiled by Charles E. Benjamin in 1985.

Descriptions of the warehouse prisons in Richmond, Va., are taken from "Andersonville: A Story of Rebel Military Prisons" by John McElroy, previously cited, chapters VII, VIII, IX, XI and XII; "Military History of Ohio", The Transcontinental Publishing Co., New York, Toledo and Chicago, 1885, pages 284 and 285; and "Andersonville Diary" by John Ransom, previously cited, chapter III.

Information on William Perry Montonye was taken from the *Mecosta County Pioneer* newspapers of Sept. 3, 1863, and July 1, 1864; "Portrait and Biographical Album of Mecosta County, Mich.", previously cited, pages 486-489; and "Alphabetical General Index to Michigan Soldiers and Sailors Individual Records", published under authority of Coleman C. Vaughan, Secretary of State, Wynkoop Hallenbeck Crawford Co. State Printers, Lansing, Mich. 1915, page 684.

Let hope Predominate, Chapter V

Descriptions of Belle Isle are taken from a personal visit on Aug. 10, 1996; "Military History of Ohio", previously cited, page 285; "Andersonville Diary" by John Ransom, previously cited, chapters I and II; "Andersonville: A Story of Rebel Military Prisons" by

John McElroy, previously cited, chapter XIII; and "Life in Rebel Prisons" by Robert H. Kellogg, previously cited, chapter X.

O, Won't We Be a Happy Crowd,
Chapter VI

The description and biographical sketch of Neal Dow is drawn from "Civil War Prisons" edited by William B. Hesseltine, Kent State University Press, Kent, Ohio, 1972, pages 60-79; and "The Civil War Dictionary", Revised Edition, by Mark Mayo Boatner III, David McKay Company Inc., New York, 1988, page 245.

Biographical information on T.M Wynn and Dan Will of Williams County, Ohio, is taken from "Official Roster of the Soldiers of the State of Ohio in the War of the Rebellion", compiled under direction of the Roster Commission, The Ohio Valley Press, Cincinnati 1888, Vol. VII., pages 419 and 422.

Quotation of John Hussey regarding conditions of the situation on Belle Isle is taken from "Official Records", Series II, Vol. VI, pages 482-483.

S.J. Radcliffe's account of the condition of sick prisoners released from Belle Isle is found in "Official Records", Series II, Vol. VI, pages 475-476.

O Dear What Shall We Do,
Chapter VII

This chapter consists entirely of Osborn Coburn's diary.

No I Shall Not Die Here,
Chapter VIII

The Confederate report of the conditions of prisoners of war in Richmond, Va., by Isaac H. Carrington appears in "Official Records", Series II, Vol. VI, pages 544-546.

Union chaplain H.C. Trumbull's account of the affairs on Belle Isle is taken from "Official Records", Series II, Vol. VI, pages 530-531.

Descriptions of the bread, meat, beans and soup issued to the prisoners were taken from "Narrative of Privations and Sufferings of United States Officers and Soldiers While Prisoners of War in the Hands of the Rebel Authorities, Being the Report of a Commission of Inquiry Appointed by the United States Sanitary Commission", printed for the U.S. Sanitary Commission by King & Baird, PRS., 607 Sansom St., Philadelphia, Pa., 1864, pages 37 and 49.

The U.S. Sanitary Commission quotation regarding the men being disabled and destroyed was taken from "Narrative of Privations and Sufferings", just cited, page 95. Quotation from Sullivan Meredith regarding retaliation against Confederate prisoners held in the North is taken from a letter that appears in "Official Records", Series II, Vol. VI, page 457-458.

To Build Castles in the Air,
Chapter IX

Biographical information on Asa Howard was

obtained from his service records and his widow's pension files in the National Archives in Washington, D.C.

It Does Require a Stout Heart,
Chapter X

The recipe for beef tea was taken from "Dr. Chase's Family Physician, Farrier, Bee-Keeper and Second Recipe Book", by Dr. A.W. Chase, Chase Publishing Co., Toledo, Ohio, 1875, page 140.

I Shall Yet Come Out All Right,
Chapter XI

Accounts of the suffering at Belle Isle due to the cold were taken from "Narrative of Privations and Sufferings of United States Officers and Soldiers While Prisoners of War", cited previously, pages 48-49, 119, 140-145, 148-150, 180, 186; "Military History of Ohio", previously cited, page 285; and "Life in Rebel Prisons" by Robert H. Kellogg, previously cited, pages 370-374.

My Old Camp Complaint,
Chapter XII

Sgt. Daniel Gribban's accounts of Belle Isle and Confederate Military General Hospital No. 21 were reported in the New York *Tribune*, and were reprinted in the Dec. 8, 1864, edition of the Bryan, Ohio, *Union Press*.

John Ransom's pea bean soup recipe is from "Andersonville Diary", previously cited, page 28.

Surgeon G.W. Semple's report on the conditions of Belle Isle is quoted from the "Official Records", Series II, Vol. VI, pages 1087 and 1088.

Nearer to an End,
Chapter XIII

The description of the Osborn Coburn's handwriting in his diary, and the Feb. 6, 1864 letter to his father are contained in "Some Pioneer Settlers of Coburntown," compiled by Charles E. Benjamin in 1985.

The quote maintaining the unfitness of the Belle Isle hospital is taken from "Military History of Ohio", previously cited, page 285.

Mortality figures cited by W.A. Carrington are taken from "Official Records", Series II, Vol. VI, page 1089.

The location of Richmond's Confederate General Hospital No. 21 was provided Aug. 10, 1995, by the staff of the Valentine Museum in Richmond from materials in the facility's collection.

Sgt. Daniel Gribban's accounts of Confederate General Hospital No. 21 were reported in the New York *Tribune*, and were reprinted in the Dec. 8, 1864, edition of the Bryan, Ohio, *Union Press*, as previously cited.

The Hospital No. 21 description by Nelson Ferguson is from "Narrative of Privations and Sufferings", previously cited, pages 166-168; the comments by De Witt Peters are on page 181.

Jonathan Wilkins' report regarding patients admitted to Confederate General Hospital No. 21 was quoted from "Official Records", Series II, Vol. VI, page 1089.

The location of Osborn Coburn's original burial and information on his reinterment in grave 584 at the Richmond National Cemetery were drawn from telephone conversations with staff members at the cemetery and from "Roll of Honor No. XIV, Names of Soldiers Who In Defence of the American Union, Suffered Martyrdom in the Prison Pens Throughout the South," Government Printing Office, Washington, D.C., 1868, page 47.

Epilogue

Information on Company I members was taken from "Record of Service of Michigan Volunteers in the Civil War 1861-1865", Vol. 36, previously cited. Sources on other individuals are previously cited.

Appendix

The review of Col. B.D. Simpson's performance at Charles Town was taken from "Official Records," previously cited, Series I, Vol. XXIX, pages 1010-1015.

Artwork

The lithographs in this book credited to Frank Leslie were taken from "Frank Leslie's Illustrated Famous Leaders and Battle Scenes of the Civil War," Mrs. Frank Leslie, Publisher, New York 1896.

Index

A

Air Line Railroad, 32, 33.
Aldrich, Dulcena, 38.
Aldrich, Eva (or Rachel), 37, 38, 58, 59, 61, 63, 66, 76, 89, 96, 98, 114, 133, 141, 148, 161, 181.
Aldrich, Thomas, 38.
amnesty, 121.
amputation, 144.
Andersonville, Ga., 74, 155-158, 160-161, 179-180, 182-183, 186.
Army of Northern Virginia, 54.
Army of the Potomac, 19, 43, 61, 113.
Ashley, J.M, 33.
Atlantic Coast, 137.
Augusta, Ga., 155, 157, 160.

B

Baltimore, Md., 130, 145.
Baton Rouge, La., 160.
Beaver Creek, 33
Beckwith, Cpl. Edwin, 155.
Belle Isle, 68-74, 78-80, 82-86, 90, 97, 101-103, 105, 113, 166, 125-126, 135-136, 139, 144-145, 151-153, 161-162, 183-184, 186-187.
Berlin, Va., 45.
Berryville, Va., 19, 24, 173-175.
Big Rapids, Mich., 23, 30-31, 34, 36, 46, 39-40, 43, 46, 58, 63, 112, 115, 151, 150.
Blacks (including niggers, Negroes), 32, 34, 43, 57, 79, 81.
Blakeslee, Schuyler E., 33.
Blue Ridge Mountains, 50, 173-174.
Bolivar, 98.
Bolivar Heights, 26.
Bossieux, Lt. Virginius, 72, 143.
Bragg, Gen. Braxton, 87, 91, 111.
Broshears, Jackson, 84.
Bryan, Ohio, 32-33, 119, 161, 186-187.
Bryan Musical Association, 33.
Bunch, Pvt. Anderson, 155.
Bunker Hill, 26.
Burdick, Theodore (bugler), 155.
Burnside, Maj. Gen. Ambrose, 57, 111-112, 129.
Butler, Maj. Gen. Benjamin, 23, 32, 128-129.

C

Camp Chase, Columbus, Ohio, 64.
Camp Kellogg, Grand Rapids, Mich., 39-40.
Canby, Brig. Gen. Ed. R.S., 178.
Carrington, Maj. Isaac, 101, 144-145, 184, 187.
Castle Thunder, 56.
Charles Town, W.Va., 18-20, 22-24,47. 53,65,128, 133,155, 159, 179,180,188; casualties at, 22, 24, 170..
Charleston, S.C., 88, 91.
Charlottesville, Va., 50.
Chattanooga, Tenn.,58, 87, 111.
Chesapeake & Ohio Canal, 29.
Christian, 79, 89, 109.
Christmas, 133.
City Point, Va., 80, 108, 115, 130.
Clark Lake, Mich., 34.
clothing, 24, 55, 80-81,86, 88-89,92, 96,pp, 101, 103, 105, 108-109, 121, 128.
Coburn, Alzada (Gay), 31.
Coburn, Betsey (Wilmot), 31.

Coburn, Edwin R., 51,58.
Coburn family, 31, 165.
Coburn, Jacob Osborn, 16-18, 22, 24, 29, 31-34, 36-47, 55, 57-59, 65, 67-68, 70, 74, 86-87, 104-105, 112-113, 134, 142-144, 148-152, 155-156, 160-161, 163-165, 179-182, 184, 186-188, 190.
Coburn, John F., 31.
Coburn, Martha "Mattie" (also Thomas, Martha), 32, 34, 48, 148, 165, 180, 190.
Coburn, Laura E., 31
Coburn, Minerva (Twadell), 31.
Coburns;' Corners, 151, 165.
Cole, Maj. Henry A., 169, 175, 183.
Columbia Township, Mich. 34.
Columbia, S.C., 103.
Columbus, Ohio, 64.
Columbus, Miss., 160.
Confederate Capitol Building, 71, 72.
Confederate commissary department, 146.
Confederate Congress, 147.
Confederate government, 63, 74, 92, 96, 98-99, 113, 115, 129.
Confederate House of Representatives, 194.
Confederate money, 75, 145.
Connolly, Cpl. James, 158.
Crawford, Pvt. John M., 158.
Culpepper, Va., 114.
Cumberland, Md., 170.
Custer, Brig. Gen. George Armstrong, 19.
Curtis, Maj. W.B., 1170.
Cutler, Pvt. Luman A., 67, 100, 156.

D

Danville, Va., 87.
David's Island, N.Y., 159.
Davis, President Jefferson, 96.
deadline, 70.
Dean, Capt. Charles W., 30, 45, 115, 133, 159,
Defiance County, Ohio, 32-33, 181.
Deits, Pvt. John, 156.
DeKalb County, Ind. 31-32, 151, 181.
Department of the Gulf, 160.
Department of West Virginia, 106, 170.
diarrhea, 31, 86, 134, 137, 146, 151-152, 155.
Dingman, Pvt. George, 125.
disease, 31, 74, 81, 86, 100, 105, 100-111, 114, 124, 137, 146, 157.
District of West Florida, 160.
Dow, Brig. Gen. Neal, 77, 79, 88, 92, 102, 160.

E

Edgerton, Ohio, 32-34, 87, 142.
8th New York Cavalry, 144.
80th New York Infantry (see also 20th New York State Militia), 113, 160.
enlistment bounty, 39.

F

Fairfax Court House, Va., 43.
Falling Waters, Md., 44.
Falmouth, Va., 43.
Fenning, Pvt. John, 158.
Fenton, Mich., 157, 159.
Ferguson, Surgeon Nelson D., 144-145, 187.
1st Connecticut Cavalry, 24.
1st Maryland Home Brigade Infantry, 170.
1st Massachusetts Heavy Artillery, 170, 190.
1st Ohio Cavalry, 147.
Fort Leavenworth, Kan., 158-159.
Fortress Monroe, 80-81.
44th Indiana Volunteer Infantry, 31.

48th New York Infantry, 125.
Franklin, Pvt. Charles, 156.
Frederick, Md., 44.
Fredericksburg, Va., 99.
Front Royal, Va., 18, 47, 173.
frostbite, 125.
Fuller, C.C., 36-37, 39, 58, 180.

G

games, 59-60, 67, 121.
Garfield, James A., 36.
General Hospital No. 21, Richmond, Va., 144, 151, 186-187.
Gettysburg, Pa., 44-45, 81, 92, 113, 160.
Gibbs, Dr., 144.
God, 59-60, 66-67, 76-77, 92, 95-96, 99, 105, 109-110, 112, 117, 124, 127, 131-133, 147-148, 161.
Gordonsville, Va., 50,52, 114.
Grand Rapids, Mich., 39, 109, 159.
Grant, Maj. Gen. Ulysses S., 57-58, 87, 111.
Gray, Col. George, 39, 41.
Greenwood, Va., 50.
Grenada, Miss., 64, 160.
Gribban, Sgt. Daniel, 135, 144 186-187.

H

Hagerstown, Md., 44.
Harpers Ferry, W.Va., 18-20, 22, 24, 45, 171-174, 176-177.
Harrisonburg, Va., 48.
Hicksville, Ohio, 32.
Hight, Sgt., 72.
Hill, Sgt., 133.
Hillsboro Gap, 45.
Hiram College, 36.
Hoag, Cpl. George C., 156.
Holly, Mich., 155-156.
Holt, Judge Advocate General Joseph, 166, 177.
Howard, Sgt. Asa C., 112-113, 115, 132, 160, 185.

Howard, Mary, 112.
Hussey, John, 79, 184.

I

Imboden, Brig. Gen. John, 19-20, 22, 23, 169, 173-175.

J

Jackson County, Mich., 34, 149, 181.
James River, 55, 68-70, 78, 97, 139..
Jarvis Hospital, Baltimore, Md., 145.
Johnson's Island Prison, Ohio, 105.

K

Kelley, Brig. Gen.. Benjamin F., 166, 171, 177.
Knoxville, Tenn., 57, 112.

L

Lake Erie, 105.
Laurent, Pvt. Moses, 158.
Lee, Gen. Robert E., 24, 44-45, 61, 93, 112, 114, 173.
Lee, Brig. Gen. W.H.F. "Rooney", 160.
Leesburg, Va., 45.
Leonard, Mich., 36-37, 39, 46, 156, 158.
Libby Prison, 52,55, 79, 160.
lice, 34,55, 62, 101-102, 130, 136, 138, 147, 180.
Limestone Station, Tenn., 160-161, 191.
Lincoln, President Abraham, 32, 110, 121.
"Listen to the Mocking Bird," 143.
Lockwood, Brig. Gen. Henry Hayes, 167.
Long Bridge, 43, 70, 72.
Longstreet, Lt. Gen. James, 112.

Lookout Mountain, 87.
Los Angeles Calif., 161.
Loudoun Heights, 173.
Lynchburg, Va., 97.

M

Mail or correspondence, 18, 29-30, 36, 40, 44, 58, 92, 96, 98, 104, 128, 133-134, 141-143, 148, 151-152, 181-182, 185, 187.
Manistee, Mich., 156=158.
Marietta, Ga., 157.
Marks, Sgt., 72.
Martinsburg, W.Va., 173-174.
Masonry, 112, 132.
Maulsby, Col. F. Wm. P., 170.
Meade, Maj. Gen. George Gordon, 61, 93, 96, 99. 112, 114.
Means, Capt. Samuel, 19.
Means Union Independent Virginia Rangers, 19.
Mecosta County, Mich., 22-23, 30, 34, 36-37, 39, 42, 112, 151, 181-183.
Mecosta County Pioneer newspaper, 29, 34, 36-37, 39-40, 42-44, 151, 181-183.
Meredith, Brig. Gen. Sullivan, 89, 104, 185.
Michigan Cavalry Brigade, 19.
Miller, Sol, 87, 160.
Miner's Artillery (see also 17th Indiana Artillery), 169.
Missionary Ridge, 103.
Mobile, Ala., 63.
Montonye, W.P., 63, 160, 183.
Moon, 1st Lt. Robert, 20-21, 30, 45-46, 65, 159, 176.
Moore, 2nd Lt. Malcom M., 129-121, 159.
Mosby, Major John., 29.
Mount Jackson, Va., 47, 49.
music, 33, 42, 143.

N

Neal, Pvt. Thomas, 22.
New Orleans, La., 128.

New Year's Day, 41, 124, 127.
New York Tribune, 135, 144, 186-187.
Nichols (first name unknown), 148.
Nicholson, Pvt. Edwin, 149.
XIX Corps, 156.
Ninth Maryland Infantry, 19-20, 166-167, 171-173.
Ninth Michigan Cavalry, 72.

O

Ohio State Legislature, 33.
Onstott, New York, 38, 58, 161.
100th Ohio Infantry, 87, 160-161.
132nd New York Infantry, 145.
Orange Court House, Va., 98-99.
Orner, Pvt. Levi, 65, 156.
Ott, Mrs., 107.
Otto, Mich., 156.

P

Paradise, Pvt. John, 157.
Parish, Pvt. Thomas, 157.
Passage Creek, 47.
Patrick, Albert S., 151.
pay, 105.
Payne, Pvt. Robert H., 111, 157.
pea bean soup, 134, 186.
Peble Island, 109-110.
Peters, De Witt, 145, 187.
Petersburg, Va., 111-112.
Pfieffer, Pvt. John, 157.
Phoenix, 50.
Point Lookout, 60.
Point of Rocks, 46.
Poolesville, Md., 43.
Port Hudson, La., 79, 160.
Potomac River, 29, 44-45.
Poynton, Mrs., 109.
prisoner deaths, 76, 81, 125, 127, 135, 137, 144-146, 151-152, 161.
prisoner escape, 26, 76, 91, 135-136, 176, 180.
prisoner exchange or parole, 22, 5759-60, 62-64, 80, 87, 89, 96, 99-

100, 104, 107-108, 113-114, 117-118, 121-123, 128-130, 142, 158, 160.

R

Radcliffe, Assistant Surgeon S.J., 80m, 184.
Raiders, 19, 74-75.
Raleigh, N.C., 95, 155.
Ransom, Sgt. John, 72, 134, 180, 183, 186.
Rappahannock River, 62.
rations, 43, 45, 48, 50, 52, 58, 60, 62-64, 75, 80, 86, 88-89, 91, 93-94, 96, 98-99, 102-108, 110-111, 114-115, 121-122, 124-125, 132-134, 136-137, 143-144, 147-148, 174.
Richmond, Va., 50, 52, 55-56, 59, 62, 64, 68, 70-73, 79, 86, 88, 91, 95, 97, 101-104, 116-117, 139, 142, 144-145, 148, 155-156, 158, 160, 163-164, 182, 184, 187-188.
Richmond National Cemetery, 151, 156, 163-164.
Richmond Whig newspaper, 117.
Riley, Pvt. Charles, 157.
Rockafellow, Lt. Benjamin "Frank", 24-25.
Rolfe, Maj. Frank A., 170.
Rosecrans, Maj. Gen. William S., 57-58.

S

Sabbath, 17, 66.
Savannah, Ga., 160-161.
2nd Maryland Potomac Home Brigade, Company F, 19.
Semple, Surgeon G. William, 136, 144, 186.
Seneca Mills, Md., 29, 44, 180.
Serrells, Mrs., 32.
17th Indiana Artillery (see also Miner's Artillery), 169.
Sharpsburg, 44.
Shaw, Pvt. William, 158.

Shenandoah River, 18, 27, 47.
Shenandoah Valley, 18, 43, 46.
Shepherdstown, Va., 159.
Simpson, Col. B.D., 20, 22, 166-178, 188.
6th Michigan Cavalry, 18-19, 22, 24, 30, 39, 43, 53, 65, 109, 111, 120, 133, 155-159, 180, 182.
Smith, Col., 41.
Smith Prison, 55.
Smith, Pvt. Walter S., 125.
Smithfield, Va., 174.
Spencer rifles, 41.
Spencer carbines, 41.
Staunton Turnpike, 27.
Staunton, Va., 50.
stealing (also see Raiders), 66, 146.
Strasburg, Va., 18.
Stronach, Mich., 155, 157-158.
Stryker, 32.

T

Teggerdine, Pvt. William, 158-159.
Thanksgiving Day, 109.
Thatcher, Pvt. Eleazar H. 157.
Third Michigan Cavalry, 63, 151, 160.
13th Maine Infantry, 160.
34th Massachusetts Infantry, 169.
Thomas, Alfred, 32.
Thomas, Ann (also Williams, Ann), 32, 58.
Cynthia D. (Harvey), 149-150, 180.
Thomas, Grantie Heman, 150.
Thomas, Hawley, 32, 36, 148-150.
Thomas, The Rev. Heman, 32, 58.
Thomas, "Ladd," 32.
Thomas, Lillian May, 150.
Thomas, Martha (also Coburn, Martha), 32, 34, 48, 148, 165, 180, 190.
tobacco, 52, 55, 66, 92, 121.

Toledo, Ohio, 32, 33, 180, 183, 186.
trading, 62, 64-67, 91, 105-108, 110, 121, 130-133, 145.
Tredegar Iron Works, 70, 116.
12th West Virginia Infantry, 170.
Tyrone, Mich, 157.
20th New York State Militia (see also 80th New York Infantry), 113, 160.
27th Michigan Infantry, 125.

U

U.S. Christian Commission, 79, 89.
U.S. government, 63, 74, 76-77, 86, 92, 96, 99-100, 104, 106-108, 113, 118, 124, 128, 132, 135, 137, 147, 174, 188.
U.S. Sanitary Commission, 82-85, 101, 103-104, 185.

V

Valentine's Day, 147.
vermin (see lice), 55, 81, 135.
Vinton, Capt. Harvey H., 45-45.

W

warehouse prison description, 55-57, 182.
Warren, Ohio, 41.
Washington, D.C., 30-31, 41, 73, 79, 179, 181, 185, 188.
Waterford, 45.
Way, Pvt. Joel B., 157.
Waynesborough County, Va., 50.
Weinrich, Pvt. Frank, 157.
Wells, Col. George D., 169, 173-175.
Western Michigan Bar, 39.
Whig, 31, 117.
White, Maj. Elijah V., 45, 173-175.
White Lake, Mich., 157, 159.
White, Mary, 112.
White, Pvt. Orrin E., 159.
Wilkins, Surgeon Jonathan, 145, 187.
Will, Pvt. Daniel, 87, 160-161, 184.
Williams County, Ohio, 33, 87, 181, 184.
Winchester, Va., 174-175.
Winder, Brig. Gen. John, 101.
Winter, James, 32.
Woodbury Musical Association, 33.
Wyndham, Sir Percy, 43.
Wynn, 1st Sgt. Turner M., 87, 161, 184.

About the Author

Don Allison, a veteran journalist and author, is a lifelong resident of Williams County, Ohio, where he shares a historic home with his wife, Diane. A 1976 graduate of Stryker High School, Don earned a bachelor of arts degree in journalism from the University of Toledo in 1980.

As a high school student Don got his start in journalism as a sports writer and photographer with the weekly Advance Reporter newspaper, now known as the Village Reporter. He joined The Bryan Times in 1981, where he served many years as news editor and recently retired as senior editor. He has received numerous Associated Press and United Press International awards for his news, feature and column writing and special section design. Even in retirement, Don's weekly column "On My Mind" continues to be a Bryan Times fixture.

Drawing on knowledge gained from a lifetime of studying the Civil War, Don has written extensively about that conflict. He and Diane are the founders of Faded Banner Publications, which publishes books on the Civil War and Northwest Ohio history, as well as the paranormal. Currently Don is co-authoring, with fellow Northwest Ohio historian Richard Cooley, a book on the 38th Ohio Volunteer Infantry in the Civil War.

For nearly four decades Don has served on the Williams County, Ohio, Historical Society Board of Trustees, and through the years he has held various offices with the organization. He spearheaded the successful effort in which the society acquired and preserved the 1845 Society of Friends Meeting House in western Williams County. Don also is a founding member and past trustee and officer of the Stryker Area Heritage Council. Currently Don serves as a historical interpreter with Sauder Village, Archbold, Ohio.

Don's other books include "I Met a Ghost at Gettysburg: A Journalist's Journey Into the Paranormal," "I Met More Ghosts at Gettysburg: A Journalist's Paranormal Journey Continues," "The Best of On My Mind: The Bryan Times Newspaper Columns of Don Allison," "The Best of On My Mind Volume II: The Bryan Times Newspaper Columns of Don Allison" and "Almost Award Winning Christmas Carols: Holiday Hilarity." Each is available for $16.95 plus $3.50 shipping and handling from Faded Banner Publications, PO Box 101, Bryan, OH 43506.

Order books online, and check out other books and products by Faded Banner Publications, at *www.fadedbanner.com.*

www.ingramcontent.com/pod-product-compliance
Lightning Source LLC
Chambersburg PA
CBHW060132100426
42744CB00007B/755